Signals and Systems
Demystified

Demystified Series

Accounting Demystified
Advanced Statistics Demystified
Algebra Demystified
Alternative Energy Demystified
Anatomy Demystified
asp.net 2.0 Demystified
Astronomy Demystified
Audio Demystified
Biology Demystified
Biotechnology Demystified
Business Calculus Demystified
Business Math Demystified
Business Statistics Demystified
C++ Demystified
Calculus Demystified
Chemistry Demystified
College Algebra Demystified
Corporate Finance Demystified
Databases Demystified
Data Structures Demystified
Differential Equations Demystified
Digital Electronics Demystified
Earth Science Demystified
Electricity Demystified
Electronics Demystified
Environmental Science Demystified
Everyday Math Demystified
Forensics Demystified
Genetics Demystified
Geometry Demystified
Home Networking Demystified
Investing Demystified
Java Demystified
JavaScript Demystified
Linear Algebra Demystified
Macroeconomics Demystified
Management Accounting Demystified

Math Proofs Demystified
Math Word Problems Demystified
Medical Billing and Coding Demystified
Medical Terminology Demystified
Meteorology Demystified
Microbiology Demystified
Microeconomics Demystified
Nanotechnology Demystified
Nurse Management Demystified
OOP Demystified
Options Demystified
Organic Chemistry Demystified
Personal Computing Demystified
Pharmacology Demystified
Physics Demystified
Physiology Demystified
Pre-Algebra Demystified
Precalculus Demystified
Probability Demystified
Project Management Demystified
Psychology Demystified
Quality Management Demystified
Quantum Mechanics Demystified
Relativity Demystified
Robotics Demystified
Signals and Systems Demystified
Six Sigma Demystified
sql Demystified
Statics and Dynamics Demystified
Statistics Demystified
Technical Math Demystified
Trigonometry Demystified
uml Demystified
Visual Basic 2005 Demystified
Visual C# 2005 Demystified
xml Demystified

Signals and Systems
Demystified

David McMahon

New York   Chicago   San Francisco   Lisbon   London   Madrid
Mexico City   Milan   New Delhi   San Juan   Seoul
Singapore   Sydney   Toronto

The McGraw·Hill Companies

Library of Congress Cataloging-in-Publication Data

McMahon, David.
   Signals and systems demystified / David McMahon.—1st ed.
         p.   cm.
   Includes index.
   ISBN 0-07-147578-8 (alk. paper)
   1. Signal processing—Mathematical models.   2. Signal processing.
   3. System analysis.   I. Title.
   TK5102.9.M398   2006
   621.382′2—dc22
                                                    2006015544

2 3 4 5 6 7 8 9 0   DOC/DOC   0 1 0 9 8

ISBN-13: 978-0-07-147578-5
ISBN-10: 0-07-147578-8

The sponsoring editor for this book was Judy Bass, the editing supervisor was David E. Fogarty, and the production supervisor was Pamela A. Pelton. It was set in Times Roman by TechBooks. The art director for the cover was Margaret Webster-Shapiro.

Printed and bound by RR Donnelley.

This book was printed on acid-free paper.

McGraw-Hill books are available at special quantity discounts to use as premiums and sales promotions, or for use in corporate training programs. For more information, please write to the Director of Special Sales, McGraw-Hill Professional, Two Penn Plaza, New York, NY 10121-2298. Or contact your local bookstore.

CONTENTS

PREFACE

Signals and Systems is a core subject in electrical engineering, and unfortunately it's one of the most difficult. Laden with heavy mathematics, many, if not all, students find courses in the areas of signal processing and systems to be very difficult. This book is aimed primarily at those students. It can serve as a supplemental text for students studying signals, systems, and communications courses in electrical engineering. The topics covered in this book are suitable for both undergraduate and graduate students.

This book is also very useful for electrical engineers who have been out of school for a long time and would like a refresher. We assume the reader has had calculus and some exposure to differential equations; however, this book is well suited for self-study. If you are an intelligent person simply looking to learn electrical engineering on your own, this is the book for you. Once you go through it you will be well prepared to tackle full-blown textbooks written on this topic.

Our approach in this text is to briefly describe concepts, theorems, and formulas and to focus on problem-solving. We explicitly demonstrate the *how-to* aspect of problem-solving. As a result each chapter is built around a core of explicitly solved problems. We try to demonstrate as many steps as possible so that the student does not have to guess how to get from Point A to Point B in a problem solution. Theorems and formulas are stated briefly. Curious students who are interested in derivations of formulas and theorems or detailed explanations of concepts can seek the references at the end of the book or their own textbook if they are interested.

Each chapter includes a chapter quiz with problems similar to those solved in the text. The answers to every question are provided at the end of the book, so that the student can try the problems and determine whether or not they have really grasped the material. A final exam and answer key at the end of the book provide the reader with a means to review the concepts laid out in the chapters.

We try to cover all major areas of signals and systems. The book begins by covering methods used to calculate energy and power in signals. Next, we spend time studying signals in the frequency domain using Fourier analysis. Other topics covered include amplitude, frequency, phase modulation, spectral analysis, convolution, the Laplace transform, and the z-transform. The primary aim of the book is to cover basic topics a student should master on a first exposure to the subject. Therefore, topics such as probability and digital signal processing are not covered in this edition.

Unfortunately, in a book with this size and scope it is not possible to cover every aspect of the field or to cover all topics in great detail. We have tried to put the best sampling of basic concepts together which is representative of most courses and texts. In any case, you should be able to get through this book relatively quickly and it will give you the confidence to solve problems and prepare you to go on to further study in the field.

David McMahon

ABOUT THE AUTHOR

David McMahon works as a researcher in the national laboratories on nuclear energy. He has advanced degrees in physics and applied mathematics, and has written several titles of McGraw-Hill.

CHAPTER 1

Introduction

A *signal* is a function of time that carries information. Physically, it may be a voltage across a capacitor or a current through a resistor for example, but in this book we will primarily be interested in the *mathematical* properties of signals. As such in general we will ignore the specific realization of a signal, beyond the understanding that it has some electrical form. It is typical to denote a signal by $x(t)$.

Often signals are real functions of time. However, it is also possible to have a signal that is a complex function. In this book we will follow the convention used in electrical engineering and denote $j = \sqrt{-1}$. Therefore a complex signal can be written in the form $x(t) = x_1(t) + jx_2(t)$, where $x_1(t)$ and $x_2(t)$ are real functions.

Continuous and Discrete Signals

A signal can be *continuous* or *discrete*. Basically, in this case the signal is a plain old function you are familiar with from calculus. In short, a continuous signal can assume any value in some continuous interval (a, b). In fact, a and b can

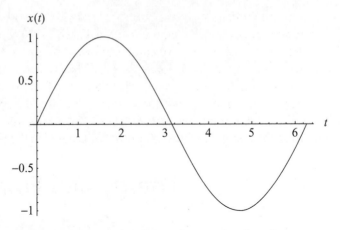

Fig. 1-1. An example of a continuous signal, a sine wave.

range over the entire real numbers, such that the signal is defined for $(-\infty, \infty)$. A continuous signal is shown in Fig. 1-1.

A *discrete time signal* is defined at discrete times we can label with an integer n. Therefore we often define a discrete time signal by $x[n]$. A discrete time signal can be created from a continuous signal by sampling $x(t)$ at regular intervals. Mathematically, we can think of a discrete signal as a *sequence* of numbers. If we denote the sampling interval by T_s, then $x[n] = x(nT_s)$; that is, we compute the discrete values of $x[n]$ by passing the argument $t = nT_s$ to the function $x(t)$. In Fig. 1-2, we show a discrete time signal formed by sampling the sin function at regular intervals.

In this chapter we will examine some basic properties of continuous signals; in the next chapter we'll do some examples with discrete signals.

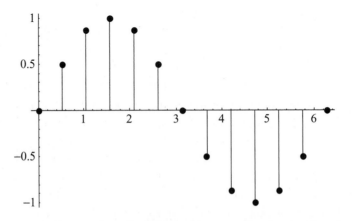

Fig. 1-2. A discrete time signal formed by sampling the sine function.

As we will see throughout much of the book, it is possible to analyze signals by studying their frequency content. Later, we will see that by using a mathematical tool known as the *Fourier transform* we can transform a function of time into a function of frequency ω, which is denoted by $X(\omega)$. For now, we will begin by looking at some important properties that can be studied as functions of time. We often say that we are working in the *time domain* when looking at signals this way. When studying $X(\omega)$, we say that we are working in the *frequency domain*. We begin by computing the energy and power content of a signal in the time domain.

Energy and Power in Signals

Consider a voltage $v(t)$ across a resistor R. Recalling Ohm's law, $v(t) = Ri(t)$, where $i(t)$ is the current, the instantaneous power is given by

$$p(t) = \frac{v(t)^2}{R}$$

Equivalently, we can instead consider the current $i(t)$ through the resistor, in which case the power is given by

$$p(t) = Ri(t)^2$$

Now if we calculate the power on a per-ohm basis, then we have $p(t) = v^2(t) = i^2(t)$. Then the total energy in joules is found by integrating

$$E = \int_{-\infty}^{\infty} p(t)\, dt = \int_{-\infty}^{\infty} i^2(t)\, dt$$

The average power is given by

$$P = \lim_{T \to \infty} \frac{1}{T} \int_{-T/2}^{T/2} i^2(t)\, dt$$

We can generalize these notions to find the energy and power content in an arbitrary signal $x(t)$. In general a signal can be complex, so we consider the squared modulus given by $|x(t)|^2 = x(t)\overline{x}(t)$, where $\overline{x}(t)$ is the complex conjugate given by $\overline{x}(t) = x_1(t) - jx_2(t)$. If the signal is real then $|x(t)|^2 = x^2(t)$.

Therefore given a signal $x(t)$, the normalized energy content E is given by

$$E = \int_{-\infty}^{\infty} |x(t)|^2\, dt \tag{1.1}$$

The normalized average power of a signal is given by

$$P = \lim_{T \to \infty} \frac{1}{T} \int_{-T/2}^{T/2} |x(t)|^2 \, dt \qquad (1.2)$$

If a signal is discrete, then the normalized energy content is found by calculating

$$E = \sum_{n=-\infty}^{\infty} |x[n]|^2 \qquad (1.3)$$

And the normalized average power becomes

$$P = \lim_{N \to \infty} \frac{1}{2N+1} \sum_{n=-N}^{N} |x[n]|^2 \qquad (1.4)$$

Classification of Energy Signals and Power Signals

Signals can be classified as *energy signals* or *power signals*. An energy signal is one for which the energy E is finite, that is $0 < E < \infty$, while the average power vanishes ($P = 0$). On the other hand, if P is nonzero but finite (i.e., $0 < P < \infty$) and the energy is infinite, then the signal is a power signal. It is possible for a signal to be neither an energy signal nor a power signal. In the next few sections, we summarize some common signal types.

DC SIGNALS

A *DC signal* is simply a signal that has a constant value. In Fig. 1-3, we show a signal that maintains the constant value of unity for all times.

PERIODIC SIGNALS

In many cases of interest a signal will be periodic. This means that there exists some positive number T_0 which we denote the *period* such that

$$x(t) = x(t + T_0) \qquad (1.5)$$

The *fundamental frequency* f_0 is given by the inverse of the period:

$$f_0 = \frac{1}{T_0} \text{ Hz} \qquad (1.6)$$

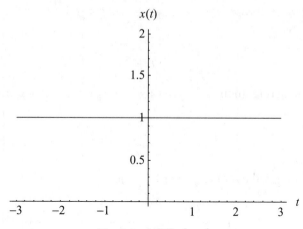

Fig. 1-3. A DC signal.

When considering periodic signals, we will examine the energy content over one period. If the energy for one period E_0 is finite, then the signal is a power signal, and the power is given by

$$P = \frac{E_0}{T_0} \quad \text{(for periodic signals)} \tag{1.7}$$

Many types of periodic functions are possible. For example, in Fig. 1-4 we have a "sawtooth" wave. To find the period, we look for the smallest value of t at which a feature of the function repeats (recall (1.5)). For example, in the sawtooth

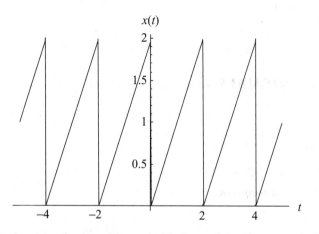

Fig. 1-4. A sawtooth wave. The period is 2 s, and the frequency is 0.5 Hz.

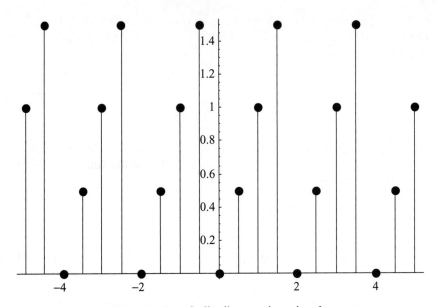

Fig. 1-5. A periodic discrete time signal.

wave we can look at the location of each peak in the wave. By inspection we see that the peaks repeat every 2 s, and so the fundamental period is $T_0 = 2$ s. The fundamental frequency is then found to be

$$f_0 = \frac{1}{T_0} = \frac{1}{2} \, \text{Hz}$$

It is also possible to have periodic discrete time signals. A discrete time signal $x[n]$ is periodic with period N, where N is a positive integer if

$$x[n] = x[n + N] \tag{1.8}$$

A simple example of a periodic discrete time signal is found by sampling the sawtooth wave at regular intervals. This is shown in Fig. 1-5.

SINUSOIDAL SIGNALS

The most familiar periodic functions are the trigonometric functions. In particular, we are interested in *sinusoidal* signals. A sinusoidal signal can be written as

$$x(t) = A \cos{(\omega t + \theta)} \tag{1.9}$$

Here A is a real number called the *amplitude* and θ is called the *phase angle*. The units of the argument are in radians. The fundamental period of a sinusoidal signal is given by

$$T_0 = \frac{2\pi}{\omega} \qquad (1.10)$$

We call ω the angular frequency. The units of angular frequency are radians per second (rad/s). Angular frequency is related to the fundamental frequency defined in (1.6) by

$$\omega = 2\pi f_0 \qquad (1.11)$$

When working with sinusoidal functions it is helpful to recall Euler's formula. In particular, we can write

$$e^{\pm j\omega t} = \cos \omega t \pm j \sin \omega t \qquad (1.12)$$

We can also manipulate this expression to write the cosine and sine functions in terms of exponentials. These formulas are

$$\cos \omega t = \frac{e^{j\omega t} + e^{-j\omega t}}{2}, \qquad \sin \omega t = \frac{e^{j\omega t} - e^{-j\omega t}}{2j} \qquad (1.13)$$

Computing Energy: Some Examples

Now let's take a step back and compute the energy and power content of a few signals. First a tip.

When computing the energy of a signal, some basic mathematical facts are useful. First recall that an *even* function of t is one that satisfies $f(-t) = f(t)$. Such a function is symmetric about the origin. A quintessential example is the cosine function, which is shown in Fig. 1-6.

We can exploit the symmetry of an even function when computing integrals. If we compute the integral of an even function over a symmetric interval a, then the following relationship holds:

$$\int_{-a}^{a} f(t)\, dt = 2 \int_{0}^{a} f(t)\, dt \qquad (1.14)$$

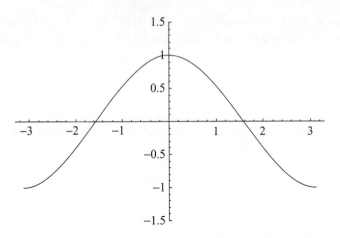

Fig. 1-6. An even function is symmetric about the origin.

This relationship also holds when integrating over the entire real line, provided that the integral in question converges

$$\int_{-\infty}^{\infty} f(t)\,dt = 2\int_{0}^{\infty} f(t)\,dt \qquad (1.15)$$

Note that $\int_{-\infty}^{\infty} \cos(t)\,dt$ does not converge.

An *odd* function is one for which $f(-t) = -f(t)$. A familiar example of an odd function is $\sin(t)$. If we plot this function, it looks "flipped over" as we cross the origin (see Fig. 1-7).

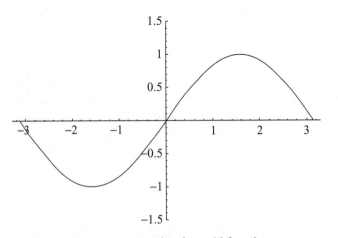

Fig. 1-7. An example of an odd function.

Looking at the plot of the sine function it is easy to see that an integral over a symmetric interval will vanish. More generally, if $f(t)$ is an odd function then

$$\int_{-a}^{a} f(t) \, dt = 0 \qquad (1.16)$$

Or, if the integral converges when integrating over the entire real line,

$$\int_{-\infty}^{\infty} f(t) \, dt = 0 \qquad (1.17)$$

Note that $\int_{-\infty}^{\infty} \sin(t) \, dt$ does not converge. We will have more to say about even and odd functions later.

Before we look at some examples, let's review some facts about energy and power signals. Consider an arbitrary signal $x(t)$. Then

- If E as defined in (1.1) is finite, then $x(t)$ is an energy signal.
- If $x(t)$ is an energy signal, then the average power $P = 0$.
- A periodic signal is a power signal.
- If the energy is E_0 over one period T_0 of a periodic signal, then the power contained in the signal is $P = E_0 / T_0$.
- A signal of finite duration is an energy signal.

EXAMPLE 1-1
Find the energy content of the exponentially decreasing signal

$$x(t) = \begin{cases} e^{-2t} & t \geq 0 \\ 0 & t < 0 \end{cases}$$

SOLUTION 1-1
We begin by computing the square

$$|x(t)|^2 = (e^{-2t})^2 = e^{-4t}$$

Using (1.1) and considering that the signal is zero for $t < 0$, we can find the energy content of the signal. We have

$$E = \int_{-\infty}^{\infty} |x(t)|^2 \, dt = \int_{0}^{\infty} e^{-4t} \, dt$$

To do the integral, we do the following substitution. Let $u = -4t$. Then we have

$$du = -4\,dt \quad \text{or} \quad dt = -\frac{1}{4}\,du$$

Now, when $t = 0$, $u = -4t = 0$. At the other limit, as $t \to \infty$, we have $u \to -\infty$. Putting all this together allows us to write

$$E = \int_0^\infty e^{-4t}\,dt = -\frac{1}{4}\int_0^{-\infty} e^u\,du$$

We can get rid of the minus sign by recalling that $\int_a^b f(t)\,dt = -\int_b^a f(t)\,dt$. From calculus we also recall that $\int e^y\,dy = e^y$, and so

$$E = -\frac{1}{4}\int_0^{-\infty} e^u\,du = \frac{1}{4}\int_{-\infty}^0 e^u\,du = \frac{1}{4}e^u\Big|_{-\infty}^0$$

$$= \frac{1}{4}(e^0 - e^{-\infty}) = \frac{1}{4}(1 - 0) = \frac{1}{4}$$

The energy is finite, and so this is an energy signal.

EXAMPLE 1-2
Consider an exponentially decreasing sinusoidal signal, given by

$$x(t) = \begin{cases} e^{-2t}\sin t & t \ge 0 \\ 0 & \text{otherwise} \end{cases}$$

Find the energy content of this signal.

SOLUTION 1-2
A plot of the function is shown in Fig. 1-8.

Once again, we start by finding the squared modulus of the function. This signal is real so we simply square it

$$|x(t)|^2 = e^{-4t}\sin^2 t$$

You will recall from calculus that to integrate the square of a sin function, it is helpful to use the trig identity

$$\sin^2 t = \frac{1 - \cos 2t}{2}$$

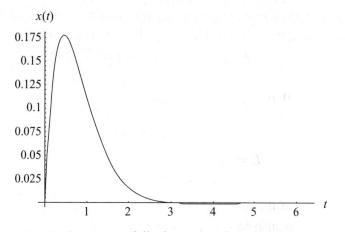

Fig. 1-8. An exponentially decreasing sinusoidal function.

Therefore we can write

$$|x(t)|^2 = e^{-4t}\sin^2 t = \frac{1}{2}e^{-4t} - \frac{1}{2}e^{-4t}\cos 2t$$

The integral to determine the energy content of the signal then becomes

$$E = \int_{-\infty}^{\infty} |x(t)|^2 dt = \int_{0}^{\infty}\left(\frac{1}{2}e^{-4t} - \frac{1}{2}e^{-4t}\cos 2t\right)dt$$

$$= \frac{1}{2}\int_{0}^{\infty} e^{-4t}dt - \frac{1}{2}\int_{0}^{\infty} e^{-4t}\cos 2t\, dt \qquad (1.18)$$

We just did the first of these integrals in Example 1-1. The first term is just

$$\frac{1}{2}\int_{0}^{\infty} e^{-4t}dt = \frac{1}{2}\left(\frac{1}{4}\right) = \frac{1}{8}$$

The second term is a bit more complicated. We will have to resort to integration by parts. Many readers will recall the technique from calculus, but let's go through the details. First, let's write down the integral

$$\int_{0}^{\infty} e^{-4t}\cos 2t\, dt$$

When faced with an exponential times a cos or sin function, integration by parts will have to be applied twice. The integration by parts formula is

$$\int u \, dv = uv - \int v \, du \tag{1.19}$$

We identify $u = \cos 2t$ and $dv = e^{-4t}$, and so

$$v = \int e^{-4t} \, dt = -\frac{1}{4} e^{-4t}$$

and $du = -2 \sin 2t \, dt$. Using (1.19) we then have

$$\int_0^\infty e^{-4t} \cos 2t \, dt = -\frac{1}{4} e^{-4t} \cos 2t \Big|_0^\infty - \frac{1}{2} \int_0^\infty e^{-4t} \sin 2t \, dt \tag{1.20}$$

Now we apply the same procedure to $\int_0^\infty e^{-4t} \sin 2t \, dt$. We take $u = \sin 2t$ and $dv = e^{-4t}$, so we again have

$$v = \int e^{-4t} \, dt = -\frac{1}{4} e^{-4t}$$

This time, $du = 2 \cos 2t \, dt$. Then the integral is

$$\int_0^\infty e^{-4t} \sin 2t \, dt = -\frac{1}{4} e^{-4t} \sin 2t \Big|_0^\infty + \frac{1}{2} \int_0^\infty e^{-4t} \cos 2t \, dt$$

The next step is to insert this result into (1.20). We obtain

$$\int_0^\infty e^{-4t} \cos 2t \, dt$$

$$= -\frac{1}{4} e^{-4t} \cos 2t \Big|_0^\infty - \frac{1}{2} \left[-\frac{1}{4} e^{-4t} \sin 2t \Big|_0^\infty + \frac{1}{2} \int_0^\infty e^{-4t} \cos 2t \, dt \right]$$

$$= -\frac{1}{4} e^{-4t} \cos 2t \Big|_0^\infty + \frac{1}{8} e^{-4t} \sin 2t \Big|_0^\infty - \frac{1}{4} \int_0^\infty e^{-4t} \cos 2t \, dt$$

Now we can move all occurrences of $\int_0^\infty e^{-4t} \cos 2t \, dt$ to the left-hand side and solve. This gives

$$\int_0^\infty e^{-4t} \cos 2t \, dt = -\frac{1}{5} e^{-4t} \cos 2t \Big|_0^\infty + \frac{1}{10} e^{-4t} \sin 2t \Big|_0^\infty$$

The constants come from adding

$$\frac{1}{4} \int_0^\infty e^{-4t} \cos 2t \, dt$$

to both sides of the relation we obtained above and then canceling the constants on the left side. Now let's evaluate the result at the limits. At $t = \infty$, the exponentials go to zero and so we don't have to consider that case. At $t = 0$ we have $\sin 2t = 0$ and $\cos 2t = 1$, and so we obtain

$$\int_0^\infty e^{-4t} \cos 2t \, dt = -\frac{1}{5} e^{-4t} \cos 2t \Big|_0^\infty + \frac{1}{10} e^{-4t} \sin 2t \Big|_0^\infty = \frac{1}{5}$$

This calculation was so long that it is possible to forget what we were doing! To compute the energy of the signal, we need to put this into (1.18). Recall that we had

$$E = \int_{-\infty}^\infty |x(t)|^2 \, dt = \frac{1}{2} \int_0^\infty e^{-4t} \, dt - \frac{1}{2} \int_0^\infty e^{-4t} \cos 2t \, dt$$

We also found that

$$\frac{1}{2} \int_0^\infty e^{-4t} \, dt = \frac{1}{8}$$

Putting everything together, the energy content of $x(t) = e^{-2t} \sin t$ defined for $t \geq 0$ is

$$E = \frac{1}{8} - \frac{1}{2}\left(\frac{1}{5}\right) = \frac{1}{8} - \frac{1}{10} = \frac{1}{40} > 0$$

The energy is finite and positive, so the signal is an energy signal. For our next example, we put a twist on the exponential function.

EXAMPLE 1-3

Consider a signal $x(t) = e^{-|t|}$. Determine the energy and power content of this signal.

SOLUTION 1-3

First we compute the squared modulus of the function

$$|x(t)|^2 = e^{-2|t|} \tag{1.21}$$

This function can be defined in the following way:

$$|x(t)|^2 = \begin{cases} e^{2t} & \text{for } t < 0 \\ e^{-2t} & \text{for } t > 0 \end{cases} \tag{1.22}$$

A plot of this function is shown in Fig. 1-9. Notice that this is an even function and therefore we can exploit the symmetry in this problem.

We could use (1.22) to split the integral into two parts, and perform the calculation this way:

$$E = \int_{-\infty}^{\infty} |x(t)|^2 \, dt = \int_{-\infty}^{0} e^{2t} \, dt + \int_{0}^{\infty} e^{-2t} \, dt$$

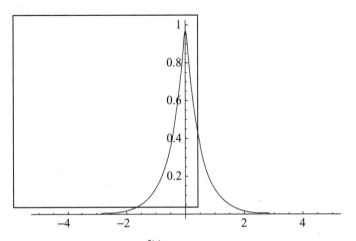

Fig. 1-9. $e^{-2|t|}$ is an even function.

However, we can simplify things given that the function is even. In this case we can use (1.15) to compute the integral. We have

$$E = \int_{-\infty}^{\infty} |x(t)|^2 \, dt = 2 \int_0^{\infty} |x(t)|^2 \, dt = 2 \int_0^{\infty} e^{-2t} \, dt$$

We can find the result by using *u-substitution*. If we let $u = -2t$ then $du = -2 \, dt$. When changing variables we must also fix up the limits of the integral. Now, when $t = 0$, we have $u = -t = 0$. When $t = \infty$, $u = -\infty$. Putting these results together we have

$$E = 2 \int_0^{\infty} e^{-t} \, dt = -\int_0^{-\infty} e^{u} \, du$$

Now let's get rid of the minus sign by recalling that

$$\int_a^b f(t) \, dt = -\int_b^a f(t) \, dt \qquad (1.23)$$

Together with the fact that $\int e^{u} \, du = e^{u}$ we have

$$E = -\int_0^{-\infty} e^{u} \, du = \int_{-\infty}^0 e^{u} \, eu = e^{u} \Big|_{-\infty}^{0} = 1$$

We obtain this result because $e^0 = 1$ and as $u \to -\infty$ we have $e^{u} \to 0$. Since the energy is finite, we conclude that this is an energy signal. We should find that the average power is zero. To find the average power, we compute

$$\frac{1}{T} \int_{-T/2}^{T/2} |x(t)|^2 \, dt$$

and then take the limit as $T \to \infty$ (recall (1.2)). Proceeding we have

$$\frac{1}{T} \int_{-T/2}^{T/2} |x(t)|^2 \, dt = \frac{1}{T} \int_{-T/2}^{T/2} e^{-2|t|} \, dt = \frac{2}{T} \int_0^{T/2} e^{-2t} \, dt$$

We have once again exploited the symmetry of the function we are integrating, and now we proceed by using the same substitution technique. This time we have

$$\frac{1}{T}\int_{-T/2}^{T/2}|x(t)|^2\mathrm{d}t = \frac{2}{T}\int_{0}^{T/2}e^{-2t}\,\mathrm{d}t = \frac{1}{T}\int_{-T}^{0}e^u\,\mathrm{d}u = \frac{1}{T}e^u\Big|_{-T}^{0} = \frac{1}{T}\left(1 - e^{-T}\right)$$

Now we take the limit:

$$P = \lim_{T\to\infty}\frac{1}{T}\left(1 - e^{-T}\right) = \lim_{T\to\infty}\frac{1}{T} - \lim_{T\to\infty}\frac{e^{-T}}{T}$$

Clearly the first term vanishes. For the second term, notice that as $T \to \infty$, $e^{-T} \to 0$. Therefore the second term vanishes as well, and we have $P = 0$ as expected for an energy signal. In the next example, we consider a periodic function and compute the energy content over a single period.

EXAMPLE 1-4
Let $x(t) = A\cos\omega t$, where A is a positive real constant. Find
   (a) the signal energy over one period
   (b) the average power of the signal

SOLUTION 1-4
   (a) The period of this signal is given by

$$T_0 = \frac{2\pi}{\omega} \tag{1.24}$$

Therefore the energy over one period is

$$E_0 = \int_{-T_0/2}^{T_0/2}|x(t)|^2\,\mathrm{d}t = \int_{-T_0/2}^{T_0/2}|A\cos\omega t|^2\,\mathrm{d}t = A^2\int_{-T_0/2}^{T_0/2}\cos^2\omega t\,\mathrm{d}t$$

To compute this integral, we recall the (hopefully) familiar trig identity

$$\cos^2 x = \frac{1 + \cos 2x}{2}$$

Using this identity the integral becomes

$$E_0 = A^2\int_{-T_0/2}^{T_0/2}\frac{1 + \cos 2\omega t}{2}\,\mathrm{d}t = \frac{A^2}{2}\int_{-T_0/2}^{T_0/2}\mathrm{d}t + \frac{A^2}{2}\int_{-T_0/2}^{T_0/2}\cos 2\omega t\,\mathrm{d}t$$

The first integral can be done readily. We find

$$\frac{A^2}{2} \int_{-T_0/2}^{T_0/2} dt = \frac{A^2}{2} t \Big|_{-T_0/2}^{T_0/2} = \frac{A^2}{2} \left(\frac{T_0}{2} + \frac{T_0}{2} \right) = \frac{A^2}{2} T_0$$

Many readers will recognize that the second integral vanishes, but let's work through the details. We can do this integral by using the u-substitution $u = 2\omega t$ from which we obtain $du = 2\omega\, dt$. This means we can write

$$\int \cos 2\omega t\, dt = \frac{1}{2\omega} \int \cos u\, du = \frac{\sin u}{2\omega}$$

Using $u = 2\omega t$ and evaluating at the endpoints we find

$$\int_{-T_0/2}^{T_0/2} \cos 2\omega t\, dt = \frac{1}{2\omega} \sin 2\omega t \Big|_{-T_0/2}^{T_0/2}$$

$$= \frac{1}{2\omega} [\sin(\omega T_0) - \sin(-\omega T_0)] = \frac{\sin(\omega T_0)}{\omega}$$

where we used the fact that $\sin(-\omega T_0) = -\sin(\omega T_0)$. Now recalling that $T_0 = 2\pi/\omega$, we have

$$\int_{-T_0/2}^{T_0/2} \cos 2\omega t\, dt = \frac{\sin(\omega T_0)}{\omega} = \frac{\sin(2\pi)}{\omega} = 0$$

Since the second integral vanishes, we have

$$E_0 = \frac{A^2}{2} T_0$$

(b) The average power for a periodic signal is easy to calculate, using (1.7). We have

$$P = \frac{E_0}{T_0} = \frac{A^2 T_0/2}{T_0} = \frac{A^2}{2}$$

EXAMPLE 1-5

Consider a trapezoidal pulse, which is defined by

$$x(t) = \begin{cases} 7 + t & -7 \leq t \leq 3 \\ 4 & -3 \leq t \leq 3 \\ 7 - t & 3 \leq t \leq 7 \end{cases}$$

(see Fig. 1-10). Find the energy content and average power of this signal.

SOLUTION 1-5

First notice that this signal is of finite duration, so it will be an energy signal and therefore $P = 0$. We could use the function definition to calculate the energy content of the signal as follows:

$$\int_{-\infty}^{\infty} |x(t)|^2 \, dt = \int_{-7}^{-3} (7 + t)^2 \, dt + 4 \int_{-3}^{3} dt + \int_{3}^{7} (7 - t)^2 \, dt$$

However, we choose to think *symmetry*. Looking at the plot in Fig. 1-4, we notice that the function is symmetric about the origin and therefore an even function. Therefore we can simplify things a bit by using (1.14). We need only consider $t \geq 0$. Specifically, we can ignore the first integral in the above expression.

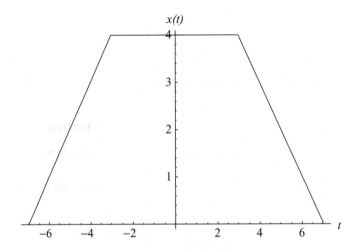

Fig. 1-10. A trapezoidal pulse.

Turning our attention to the middle piece, we have

$$4 \int_{-3}^{3} dt = 4(2) \int_{0}^{3} dt = 8t \Big|_{0}^{3} = 24$$

Next we consider the last integral, $\int_{3}^{7} (7 - t)^2 \, dt$. We use the substitution $u = 7 - t \Rightarrow du = -dt$. Looking at the limits of the integral, when $t = 3$ we have $u = 7 - 3 = 4$. At the upper limit, $t = 7$ and so $u = 7 - 7 = 0$. Therefore we find

$$\int_{3}^{7} (7 - t)^2 \, dt = - \int_{4}^{0} u^2 \, du = \int_{0}^{4} u^2 \, du = \frac{1}{3} u^3 \Big|_{0}^{4} = \frac{4^3}{3} = \frac{64}{3}$$

In order to account for the symmetry of the problem, we will double this result. The energy content of the signal is then

$$E = 24 + \frac{128}{3} = \frac{72}{3} + \frac{128}{3} = \frac{200}{3}$$

More on Even and Odd Functions

The *even part* of a function, which we denote $x_e(t)$, can be constructed from any function $x(t)$ via

$$x_e(t) = \frac{x(t) + x(-t)}{2} \qquad (1.25)$$

We can also construct the odd part $x_0(t)$ by writing

$$x_0(t) = \frac{x(t) - x(-t)}{2} \qquad (1.26)$$

Even and odd functions also have the following properties:

- The product of two even functions is even, since $x_1(-t)x_2(-t) = x_1(t)x_2(t)$.
- The product of two odd functions is even, since $x_1(-t)x_2(-t) = [-x_1(t)][-x_2(t)] = x_1(t)x_2(t)$.
- The product of an even function times an odd function is odd. If x_1 is even and x_2 is odd, then $x_1(-t)x_2(-t) = x_1(t)x_2(-t) = x_1(-t)[-x_2(-t)] = -x_1(t)x_2(t)$.

EXAMPLE 1-6
Find the even and odd components of

$$x(t) = 2 \cos t - \sin t + 3 \sin t \cos t$$

SOLUTION 1-6
We can find the even component using (1.25). First we find $x(-t)$, relying on the fact that $\cos(-t) = \cos(t)$ and $\sin(-t) = -\sin(t)$:

$$x(-t) = 2 \cos(-t) - \sin(-t) + 3 \sin(-t) \cos(-t)$$
$$= 2 \cos t + \sin t - 3 \sin t \cos t$$

Therefore the even component is

$$x_e(t) = \frac{x(t) + x(-t)}{2} = \frac{4 \cos t}{2} = 2 \cos t$$

The odd component is calculated using (1.26). We have

$$x_0(t) = \frac{x(t) - x(-t)}{2} = \frac{-2 \sin t + 6 \sin t \cos t}{2} = -\sin t + 3 \sin t \cos t$$

EXAMPLE 1-7
Determine whether or not the following functions are even or odd:

(a) $x(t) = t \cos t$
(b) $x(t) = \cos t \sin^2 t$
(c) $x(t) = t \sin t$

SOLUTION 1-7
(a) We have an odd function t multiplied by an even function $\cos t$. The product of an odd function with an even function is odd, so in this case $x(t)$ is odd.
(b) Let's look at this one in two steps. First, we note that $\cos t$ is even. Now $\sin t$ is odd, and we have $\sin^2 t = \sin t \sin t$, which is an odd function times an odd function. An odd function times an odd function is even, and so $\sin^2 t$ is even. In the end we have $x(t) = \cos t \sin^2 t$, which is an even function times an even function, so this signal is even.
(c) In this case we have an odd function times an odd function. We have $x(-t) = -t \sin(-t) = t \sin t$, so this is an even function.

Quiz

1. Suppose that $x(t) = u(t)$, where $u(t)$ is the unit step function, defined by

$$u(t) = \begin{cases} 0 & t < 0 \\ 1 & t > 0 \end{cases}$$

   Find the energy and power content of this signal. Is it an energy signal or power signal?

2. Suppose that a signal is given by

$$x(t) = \cosh(t)\,[u(t+1) - u(t-1)]$$

   A plot of this function is shown in Fig. 1-11. Find the energy content of this signal.

3. Find the energy and power content of the triangular pulse shown in Fig. 1-12.

4. Suppose that $x(t) = t\,\cos \omega t$. Find the energy content over one period.

5. For the signal in the previous problem, find the power content of the signal.

6. Find the even and odd components of $x(t) = \cos t\,\sin^2 t + \sin t\,\cos^2 t$.

7. Is the following function even or odd?

$$x(t) = t^3\,\sin^2 t\,\cos t$$

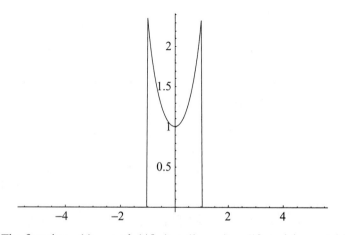

Fig. 1-11. The function $x(t) = \cosh(t)[u(t+1) - u(t-1)]$ vanishes outside $|x| > 1$.

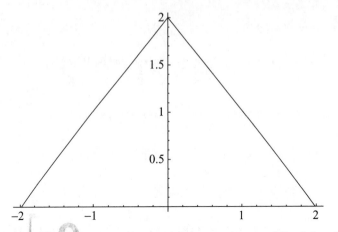

Fig. 1-12. A triangular pulse. Left of the origin $x(t) = 2 + t$, while right of the origin $x(t) = 2 - t$. The function is zero otherwise.

8. Find the energy content of

$$x(t) = \begin{cases} e^{-t} & \sin 3t \quad t \geq 0 \\ 0 & \text{otherwise} \end{cases}$$

9. What is the fundamental period for a sinusoidal signal?
10. Consider a discrete time signal. What equation does a signal satisfy if it is periodic?

CHAPTER 2

Linear Time-Invariant Systems

In this chapter we consider some basic properties of systems that are important in signal analysis. We begin by defining the term *system*. A system is a mathematical model that represents the transformation of some input signal $x(t)$ into an output signal $y(t)$. This is shown schematically in Fig. 2-1.

Physically, the system could be some circuit for example. A simple system we could imagine is that the "system" is a resistor. Using Ohm's law $v(t) = Ri(t)$, we see that the system relationship is $y(t) = Rx(t)$.

For our purposes, we aren't going to worry about the particulars of how the system is implemented. In effect it will be a black box and our only concern will be the mathematical properties of the system. Mathematically, we represent the system by a *transformation* or *operator* that we will denote by $\hat{T}$. Then we can write the action of a system on an input signal as

$$y(t) = \hat{T}\{x(t)\} \tag{2.1}$$

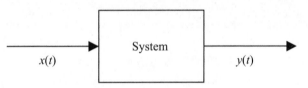

Fig. 2-1. A schematic representation of a system. The system transforms the input signal $x(t)$ into some output signal $y(t)$.

In the case of the resistor, $\hat{T}$ is simply multiplication by the constant R. Mathematically, a transformation or operator is simply an instruction to *do something* to the input function it acts on. Another simple but more complicated example than the resistor is the relationship between the current and voltage of a capacitor. In that case, we have

$$i(t) = C\frac{dv}{dt}$$

So the transformation in this case is the derivative operator $\hat{T} = d/dt$ multiplied by the constant C.

A system is known as a *continuous time system* if $x = x(t)$ and $y = y(t)$ are continuous time signals. It is also possible to have systems with input and output signals that are discrete time signals; we will discuss them in the next chapter. For now, let's go through some basic properties of systems that are important.

Memoryless Systems

If the output $y(t)$ of a given system depends only on the input $x(t)$ at the *same time*, then the system is called *memoryless*. A simple example of a memoryless system is an output that depends on a constant multiple of the input

$$y(t) = \alpha x(t)$$

where α is some real constant.

Systems with Memory

A system with memory is one where the output depends on the values of the input at previous times. An example of such a system is one where we add up (integrate) the values of the input signal from all past times up to the present time; i.e.,

$$y(t) = \int_{-\infty}^{t} x(\tau)\,d\tau$$

EXAMPLE 2-1
Determine if the following systems are memoryless:
  (a) $y(t) = \sin(t)\cos(t)$
  (b) $y(t) = \int_{-\infty}^{t/3} x(\tau)\,d\tau$ for some general input function $x(t)$
  (c) $y(t) = \int_{t_0}^{t} x(\tau)\,d\tau$, consider $x(t) = t\,e^{-t}$

SOLUTION 2-1
  (a) This system is memoryless, since the output $y(t)$ depends only on the present value of the input $x(t)$.
  (b) This system has memory, because it depends on all past values of the input.
  (c) Let's do the integral to see what the output function is. We have

$$y(t) = \int_{t_0}^{t} \tau e^{-\tau}\,d\tau$$

We can do the integral using integration by parts. We take $u = \tau \Rightarrow du = d\tau$ and $dv = e^{-\tau} \Rightarrow v = -e^{-\tau}$. Application of the integration by parts formula in this case gives

$$uv - \int v\,du = -\tau e^{-\tau}\Big|_{t_0}^{t} + \int_{t_0}^{t} e^{-\tau}\,d\tau = -\tau e^{-\tau} - e^{-\tau}\Big|_{t_0}^{t}$$

$$= -te^{-t} + t_0 e^{-t_0} - e^{-t} + e^{-t_0}$$

And so we have

$$y(t) = e^{-t_0}(1 + t_0) - e^{-t}(1 + t)$$

This system is *not* memoryless (i.e., has memory), because the output depends on the present time t and the previous time t_0.

Causal and Noncausal Systems

If a system output $y(t)$ depends only on the input at present or earlier times, we say that the system is *causal*. Another way to say this is that the output does not anticipate *future* values of the input. Any real time-dependent system is causal because the laws of physics demand it. Time moves only in the forward direction and causes precede effects in time ordering. All memoryless systems are causal.

A noncausal system anticipates the future values of the input signal in some way. For example

$$y(t) = Cx(t + a)$$

where a is a real constant, is a noncausal system.

Linear Systems

An operator $\hat{T}$ is called *linear* if the following relationship holds. Suppose that $\hat{T}$ acts on two input signals $x_1(t)$ and $x_2(t)$ to produce output signals

$$\hat{T}\{x_1(t)\} = y_1(t) \quad \text{and} \quad \hat{T}\{x_2(t)\} = y_2(t)$$

Now let a, b be two constants. The transformation $\hat{T}$ is linear if

$$\hat{T}\{ax_1(t) + bx_2(t)\} = ay_1(t) + by_2(t) \tag{2.2}$$

If a system is represented mathematically by a linear transformation, we say that the system is a *linear system*. To determine if a system is linear, use the following steps:

- Consider two input-output relationships $y_1(t)$ and $y_2(t)$ and form the sum $ay_1(t) + by_2(t)$.
- Supply the sum $ax_1(t) + bx_2(t)$ to the system and construct $\hat{T}\{ax_1(t) + bx_2(t)\}$. If this is equal to $ay_1(t) + by_2(t)$ for scalars a, b then the system is linear.

EXAMPLE 2-2
Determine if the following systems are linear:

(a) $y(t) = Rx(t)$, where R is a constant

(b) $y(t) = x^3(t)$

(c) $y(t) = d^2x/dt^2$

SOLUTION 2-2

(a) We consider two input and output signals multiplied by scalars. Since $y(t) = Rx(t)$, we have

$$y_1(t) = Rx_1(t), \qquad y_2(t) = Rx_2(t)$$

And so the sum weighted by two constants is

$$ay_1(t) + by_2(t) = aRx_1(t) + bRx_2(t)$$

Now we use the sum $ax_1(t) + bx_2(t)$ as input to the system, where $\hat{T}\{x(t)\} = Rx(t)$. We have

$$\hat{T}\{ax_1(t) + bx_2(t)\} = R[ax_1(t) + bx_2(t)] = aRx_1(t) + bRx_2(t)$$

Since this is equal to $ay_1(t) + by_2(t)$, we conclude the system is linear.

(b) For $y(t) = x^3(t)$ we have

$$ay_1(t) + by_2(t) = ax_1^3(t) + bx_2^3(t)$$

Now consider the transformation acting on $ax_1(t) + bx_2(t)$. This gives

$$\hat{T}\{ax_1(t) + bx_2(t)\} = [ax_1(t) + bx_2(t)]^3$$

As a quick aside, an easy way to obtain the coefficients in a binomial expansion is to use Pascal's triangle, the first few lines of which are

$$1$$
$$1\ 1$$
$$1\ 2\ 1$$
$$1\ 3\ 3\ 1$$
$$1\ 4\ 6\ 4\ 1$$
$$\cdots$$

In the case of the cubic, we have

$$(x + y)^3 = x^3 + 3x^2y + 3xy^2 + y^3$$

And so the transformation on the sum is

$$\hat{T}\{ax_1(t) + bx_2(t)\} = [ax_1(t) + bx_2(t)]^3$$
$$= a^3 x_1^3(t) + 3a^2 b x_1^2(t)x_2(t)$$
$$+ 3ab^2 x_1(t)x_2^2(t) + b^3 x_2(t)$$

Clearly, $\hat{T}\{ax_1(t) + bx_2(t)\} \neq ay_1(t) + by_2(t)$, so the system is not linear.

(c) For the final transformation, we have

$$ay_1(t) + by_2(t) = a\frac{d^2 x_1}{dt^2} + b\frac{d^2 x_2}{dt^2}$$

Now,

$$\hat{T}\{ax_1(t) + bx_2(t)\} = \frac{d^2}{dt^2}[ax_1(t) + bx_2(t)] = \frac{d^2}{dt^2}[ax_1(t)]$$
$$+ \frac{d^2}{dt^2}[bx_2(t)] = a\frac{d^2 x_1}{dt^2} + b\frac{d^2 x_2}{dt^2}$$

In this case, $\hat{T}\{ax_1(t) + bx_2(t)\} = ay_1(t) + by_2(t)$ and so the transformation is linear.

Time-Invariance

Consider some time-shift in the input signal, i.e.,

$$x(t) \rightarrow x(t \pm \tau)$$

The time-shift can be a delay or an advance. If such a time-shift in the input signal causes the *same* time-shift in the output signal, we say that the system is *time-invariant*. A linear system that is also time-invariant is known as a *linear time-invariant system* or LTI. For a given system with transformation $\hat{T}$, we write

$$y_\tau(t) = \hat{T}\{x(t - \tau)\}$$

If $y_\tau(t) = y(t - \tau)$ then the system is time-invariant. That is, apply a time-shift to the input, apply the transformation to obtain an output, and if this is equal to the time-shift of the original output signal, the system is time-invariant. We will illustrate this below with examples so don't worry if you are confused.

System Stability

We say that a signal $x(t)$ is *bounded* if we can find a constant α such that for all t

$$\left|x(t)\right| \leq \alpha \tag{2.3}$$

If the output signal $y(t)$ is also bounded, that is given $y(t) = \hat{T}\{x(t)\}$ and some constant β we have

$$\left|y(t)\right| \leq \beta$$

then we say that the system is *bounded-input, bounded-output stable* or BIBO.

EXAMPLE 2-3
Determine if each of the following systems are memoryless, causal, stable, and time-invariant.
  (a) $y(t) = \sin\left[x(t)\right]$
  (b) $y(t) = 2e^{-4\cos[x(t)]}$ for $t \geq 0$, 0 otherwise
  (c) $y(t) = x(t - 6)$
  (d) $y(t) = x^2(t)$
  (e) $y(t) = x(t)\cos t$
  (f) $y(t) = dx/dt$
  (g) $y(t) = \int_\infty^{t/3} x(s)\,ds$

SOLUTION 2-3
  (a) $y(t) = \sin\left[x(t)\right]$ is memoryless since the output depends only on the input at the present time. A memoryless system is causal, so $y(t)$ is causal because it does not anticipate any future values of the input and depends only on the present time.
      Next we investigate stability. The system $y(t) = \sin\left[x(t)\right]$ is stable because we know that $\left|\sin(\tau)\right| \leq 1$ for all values of τ, and so the system is bounded.

Finally, we check time-invariance. First we time-shift the input to give $x(t - \tau)$. Then we obtain $y_\tau(t) = \sin\left[x(t - \tau)\right]$. This is exactly what we obtain for $y(t - \tau) = \sin\left[x(t - \tau)\right]$, so the system is time-invariant.

(b) First, we look at a plot of the function. For an example, we choose $x(t) = t$, shown in Fig. 2-2 .

The system is memoryless, since it depends only on the input at the present time t. Since it is memoryless it is also causal.

Is the system stable? The system attains a maximum at $2e^4 \approx 109$ and so $\left|y(t)\right| \le 110$ for all t. Try playing around with different functions $x(t)$ for input. You will see that if $x(t)$ is bounded, so is $y(t)$ (in fact try examining some unbounded functions; remember that $|\cos t| \le 1$ for any value of t). So, the system is bounded, and therefore the system is stable.

The system is time-invariant since $y(t - \tau) = 2e^{-4\cos[x(t-\tau)]}$ and given input $x(t - \tau)$, $\hat{T}\{x(t - \tau)\} = 2e^{-4\cos[x(t-\tau)]}$ for any t and τ.

(c) Next we consider $y(t) = x(t - 6)$. The system is not memoryless, because it depends on the value of the input at other values than the present time t. However, the system is causal, because it depends only on previous values of t.

The system is stable if $x(t)$ is bounded by some constant α, then $\left|x(t)\right| \le \alpha \Rightarrow \left|y(t)\right| \le \alpha$.

Is the system time-invariant? First we note that the time-shifted output is $y(t - \tau) = x(t - \tau - 6)$.

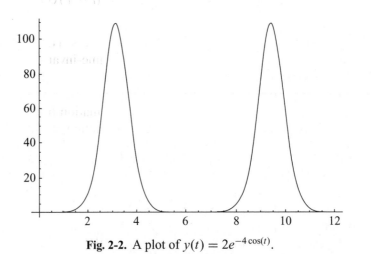

Fig. 2-2. A plot of $y(t) = 2e^{-4\cos(t)}$.

We suppose that $x_\tau(t) = x(t - \tau)$ is the time-shifted input. It is a simple matter to see that

$$y_\tau(t) = \hat{T}\{x_\tau(t)\} = \hat{T}\{x(t - \tau)\} = x(t - \tau - 6) = y(t - \tau).$$

Therefore the system is time-invariant.

(d) The system $y(t) = x^2(t)$ is memoryless since it depends only on the present time, and therefore it is causal. The system is stable if $x(t)$ is bounded by some constant α, since $|x(t)| \leq \alpha \Rightarrow |y(t)| \leq \alpha$.

Finally, we examine time-invariance. Given a time-shifted input $x(t - \tau)$, we obtain

$$y_\tau(t) = \hat{T}\{x(t - \tau)\} = x^2(t - \tau)$$

Since $y(t - \tau) = x^2(t - \tau) = y_\tau(t)$, the system is time-invariant.

(e) The system described by $y(t) = x(t)\cos t$ is memoryless, since it depends only on the present time t. Since it is memoryless, it is causal.

The system is not stable, since even if $|x(t)| \leq \alpha$, the hyperbolic cosine function is unbounded for large values of t. So $y(t)$ grows without bound and the system is not stable.

Now let $x_\tau(t) = x(t - \tau)$ be the time-shifted input. Applying the transformation to the time-shifted input we obtain

$$y_\tau(t) = \hat{T}\{x(t - \tau)\} = x(t - \tau)\cos t$$

However, a time-shift of the output gives $y(t - \tau) = x(t - \tau) \times \cosh(t - \tau) \neq y_\tau(t)$, so the system is not time-invariant.

(f) The system $y(t) = dx/dt$ is not memoryless, because the derivative of a function cannot be determined from knowledge of the function at a single point alone (therefore we need information from past times). The system is causal, since the output does not anticipate future values of the input.

The system is not stable. Consider $x(t) = (2 - t)^{2/3}$ defined for $0 \leq t \leq 2$ (see Fig. 2-3). Then

$$\frac{dx}{dt} = -\frac{2t}{3(2 - t)^{1/3}}$$

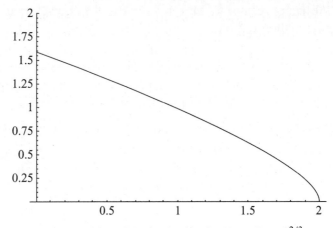

Fig. 2-3. A plot of the input signal $x(t) = (2 - t)^{2/3}$.

Although $x(t) \to 0$ as $t \to 2$, $y(t) \to -\infty$. Since we have found an example where $y(t)$ is not bounded even though the input signal is, the system is not stable.

Let's consider time-invariance. Since $y(u) = dx(u)/du$ for some arbitrary variable u, for the time-shift we have

$$y(t - \tau) = \frac{dx\,(t - \tau)}{d\,(t - \tau)}$$

Now consider the transformation of the time-shifted input. For clarity, let $u = t - \tau$. Then

$$\hat{T}\{x(t - \tau)\} = \frac{d}{dt}\big[x(t - \tau)\big] = \frac{du}{du}\frac{dx(t - \tau)}{dt} = \frac{du}{dt}\frac{dx(t - \tau)}{du}$$

But we have

$$\frac{du}{dt} = \frac{d\,(t - \tau)}{dt} = 1$$

Using this fact together with $u = t - \tau$, we see that

$$\hat{T}\{x(t - \tau)\} = \frac{dx(t - \tau)}{d\,(t - \tau)} = y(t - \tau)$$

So the system is time-invariant.

(g) The system described by $y(t) = \int_{-\infty}^{t/3} x(s)\,ds$ is not memoryless, since it depends on the values of the input at previous times. It is causal, since it does not depend on future values of the input function. To find the output, we have to add up the values of the input signal for all times in the past up to $t/3$. But $t/3$ is less than the present time t, so the output does not depend on future values of the input. Hence it is causal.

Next we investigate stability. For a counter example, suppose that the input is given by the unit step function $x = u(t)$. Then the integral becomes

$$y(t) = \int_{-\infty}^{t/3} x(s)\,ds = \int_{-\infty}^{t/3} u(s)\,ds = \int_{0}^{t/3} ds = \frac{1}{3}t$$

As shown in Fig. 2-4, clearly this function grows without bound, even though the input $|x(t)| = 1$ for all times, and so the system is not stable.

To investigate time-invariance, note that

$$y(t - \tau) = \int_{-\infty}^{(t-\tau)/3} x(s)\,ds$$

Given an input $x(t - \tau)$, the system produces the output

$$\hat{T}\{x(t - \tau)\} = \int_{-\infty}^{t/3} x(s - \tau)\,ds$$

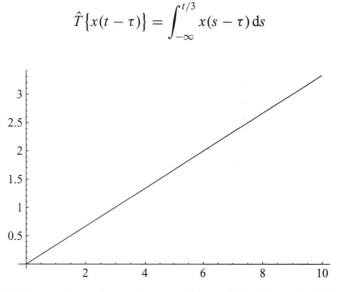

Fig. 2-4. For a unit-step input, the output in part (g) of Example 2-3 grows without bound.

Let $u = s - \tau$. Then $du = ds$. At the lower limit for $s = -\infty$, $u = -\infty$ and at the upper limit for $s = t/3$, $u = s - \tau = t/3 - \tau$, and so the integral becomes

$$\hat{T}\{x(t - \tau)\} = \int_{-\infty}^{t/3} x(s - \tau)\,ds = \int_{-\infty}^{t/3-\tau} x(u)\,du$$

$$= \int_{-\infty}^{(t-3\tau)/3} x(u)\,du = y(t - 3\tau) \neq y(t - \tau)$$

Hence, the system is not time-invariant.

The Unit Impulse Function

In this section we review the Dirac delta or *unit impulse* function $\delta(t)$. We aren't going to "derive" or justify this function in this book, suffice it to say that the unit impulse is an infinitely high spike located at a single point. For $\delta(t)$, the "spike" is located at the origin, as shown in Fig. 2-5.

We can basically define the unit impulse by writing

$$\delta(t) = \begin{cases} 0 & t \neq 0 \\ \infty & t = 0 \end{cases} \tag{2.4}$$

We can shift the position of the unit impulse by changing the argument. To move it to the right, we let $t \to t - a$, where a is some constant. In Fig. 2-6, we see $\delta(t - 2)$.

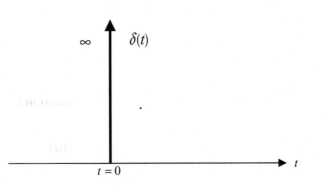

Fig. 2-5. A unit impulse located at the origin. The "spike" is located at the point $t = 0$ and its height goes to infinity.

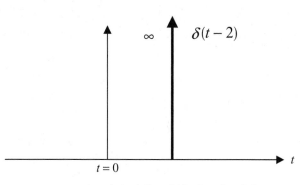

Fig. 2-6. A unit impulse shifted to the right.

We can also shift a unit impulse to the left by considering $\delta(t + a)$. The first property of interest is that the integral of the unit impulse function is unity. That is,

$$\int_{-\infty}^{\infty} \delta(t)\,dt = 1 \tag{2.5}$$

This is also true if the unit impulse is shifted left or right; i.e.,

$$\int_{-\infty}^{\infty} \delta(t \pm a)\,dt = 1 \tag{2.6}$$

The fact that the unit impulse function picks out a single point along the real line gives rise to an important property of use in signal processing, the *sampling property*. Given some function $\phi(t)$, when we integrate, a unit impulse at the origin picks out the value of $\phi(t)$ at the origin

$$\int_{-\infty}^{\infty} \phi(t)\delta(t)\,dt = \phi(0) \tag{2.7}$$

A shifted unit impulse picks out the value of the function at $t = a$

$$\int_{-\infty}^{\infty} \phi(t)\delta(t - a)\,dt = \phi(a) \tag{2.8}$$

If the point where the unit impulse is located is not included in the range of integration, then the result is zero. For example, if the unit impulse is at the origin, then

$$\int_2^{10} f(t)\delta(t)\,dt = 0$$

While we have

$$\int_2^{10} f(t)\delta(t-3)\,dt = f(3)$$

we see here that the unit impulse has picked out the value $f(3)$, illustrating the sampling property. More properties of the unit impulse that are useful to remember are

$$\delta(at) = \frac{1}{|a|}\delta(t) \qquad (2.9)$$

$$\delta(-t) = \delta(t) \qquad (2.10)$$

$$x(t)\delta(t) = x(0)\delta(t) \qquad (2.11)$$

$$x(t)\delta(t-\tau) = x(\tau)\delta(t-\tau) \qquad (2.12)$$

$$\int_{-\infty}^{\infty} \phi(t)\delta'(t-a)\,dt = -\phi'(a) \qquad (2.13)$$

An important observation is that any continuous time signal can be written in terms of the unit impulse function as

$$x(t) = \int_{-\infty}^{\infty} x(\tau)\delta(t-\tau)\,d\tau \qquad (2.14)$$

The Unit Step Function

The unit step function is denoted by $u(t)$ and defined as

$$u(t) = \begin{cases} 1 & t > 0 \\ 0 & t < 0 \end{cases} \qquad (2.15)$$

In Fig. 2-7, we show a plot of the unit step function.

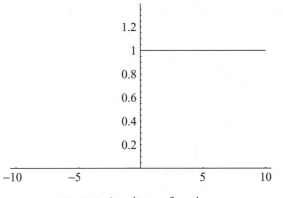

Fig. 2-7. A unit step function.

The unit step function can be shifted to the right by writing

$$u(t - a) = \begin{cases} 1 & t > a \\ 0 & t < a \end{cases} \tag{2.16}$$

An example is shown in Fig. 2-8.

When the unit step function is present in an integral, it can change the limits of integration, since it is zero for many values. For example,

$$\int_{-\infty}^{\infty} u(t)x(t)\, dt = \int_{0}^{\infty} x(t)\, dt$$

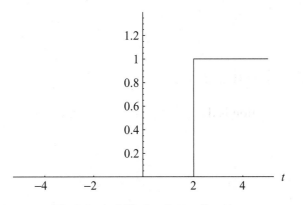

Fig. 2-8. A shifted unit step function.

Or for a shifted unit step function

$$\int_{-\infty}^{\infty} u(t-3)x(t)\,dt = \int_{3}^{\infty} x(t)\,dt$$

We can create a square pulse by adding together and subtracting unit step functions. First, notice that

$$-u(t) = \begin{cases} -1 & t > 0 \\ 0 & t < 0 \end{cases}$$

This is shown in Fig. 2-9.

To see how to construct a square pulse, look at the plots in Figs. 2-7 and 2-8. Notice that for $0 < t < 2$ we have $u(t) = 1$ but $u(t-2) = 0$. So if we compute the difference, in this region $u(t) - u(t-2) = 1$. For $t > 2$, $u(t) = 1$ and $u(t-2) = 1$, so the difference is $u(t) - u(t-2) = 0$. This gives us a square pulse. A plot is shown in Fig. 2-10.

Since a square pulse is zero outside of a given range, it simplifies integrals. Consider the square pulse $u(t-1) - u(t-2)$ shown in Fig. 2-11.

Consider the product $\left[u(t-1) - u(t-2)\right]\cos t$, which picks out values of $\cos t$ for $1 < t < 2$. This is shown in Fig. 2-12.

Notice that

$$\int_{-\infty}^{\infty} \left[u(t-1) - u(t-2)\right]\cos t\,dt = \int_{1}^{2} \cos t\,dt = \sin 2 - \sin 1$$

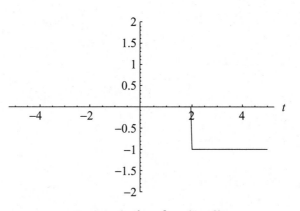

Fig. 2-9. A plot of $-u(t-2)$.

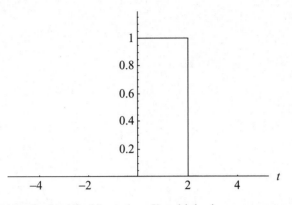

Fig. 2-10. A plot of $u(t) - u(t - 2)$, which gives a square pulse.

The unit impulse function and the unit step function are related via the derivative

$$\delta(t) = \frac{\mathrm{d}u(t)}{\mathrm{d}t} \tag{2.17}$$

Conversely, we can express the unit step function in terms of the unit impulse function via

$$u(t) = \int_{-\infty}^{t} \delta(\tau)\,\mathrm{d}\tau \tag{2.18}$$

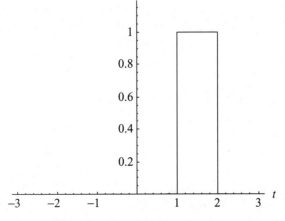

Fig. 2-11. A square pulse given by $u(t - 1) - u(t - 2)$.

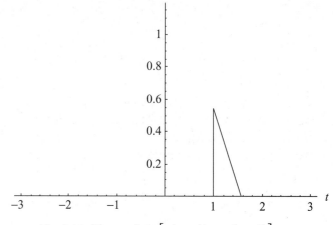

Fig. 2-12. The product $[u(t-1) - u(t-2)]\cos t$.

Impulse Response of an LTI

When an input signal to an LTI system $\hat{T}$ is the unit impulse function, we obtain the *impulse response* of the system. The impulse response is denoted by $h(t)$, so we write

$$h(t) = \hat{T}\{\delta(t)\} \tag{2.19}$$

The impulse response of a system can be used to determine the response of the system to any arbitrary input. First, we recall from (2.14) that we can represent any signal $x(t)$ in terms of the unit impulse function. Since we are considering linear systems only at this point, we can write

$$y(t) = \hat{T}\{x(t)\} = \hat{T}\left\{\int_{-\infty}^{\infty} x(\tau)\delta(t-\tau)\,d\tau\right\} = \int_{-\infty}^{\infty} x(\tau)\hat{T}\{\delta(t-\tau)\}\,d\tau$$

From (2.19), we already know the response of the system to the unit impulse function. In the case of interest here, the fact that the unit impulse function inside the integral is shifted in time makes no difference—we are studying time-invariant systems. Therefore we can extend (2.19) and write $h(t-\tau) = \hat{T}\{\delta(t-\tau)\}$. We conclude that the response of an LTI to any arbitrary input

$x(t)$ is given by

$$y(t) = \int_{-\infty}^{\infty} x(\tau)h(t-\tau)\,d\tau \qquad (2.20)$$

Many readers will recognize that (2.20) is the convolution of the input signal $x(t)$ with the impulse response $h(t-\tau)$. Therefore, we say that the output of an LTI system is given by the convolution of the input signal with the impulse response $h(t)$. We will explore convolution in more detail below.

System Step Response

As we followed the discussion of the unit impulse function by the unit step function, we now consider the response of a system to a unit step input. This is called the *step response* and denoted by $s(t)$. Specifically

$$s(t) = \hat{T}\{u(t)\} \qquad (2.21)$$

The step response can be determined by recalling the main result of the last section—that the response of an LTI to any input signal is found by computing the convolution of that signal with the impulse response of the system. Therefore we can write

$$s(t) = u(t)^*h(t) = \int_{-\infty}^{\infty} u(\tau)h(t-\tau)\,d\tau$$

In a later section where we discuss convolution in detail, we will see that convolution is commutative, meaning that $u(t)^*h(t) = h(t)^*u(t)$. It is more convenient to write the step response in the following way:

$$s(t) = h(t)^*u(t) = \int_{-\infty}^{\infty} h(\tau)u(t-\tau)\,d\tau = \int_{-\infty}^{t} h(\tau)\,d\tau \qquad (2.22)$$

Looking at (2.22), notice that if we know the step response of a system, we can find the impulse response by computing

$$h(t) = \frac{ds}{dt} \qquad (2.23)$$

This is not surprising given (2.17), which relates the unit impulse function to the unit step function.

Impulse Response and System Properties

By examining the impulse response $h(t)$ for a system, we can determine whether or not an LTI system is memoryless, causal, and stable. For an LTI system a memoryless system is one for which

$$y(t) = Kx(t)$$

where K is a constant called the gain constant. In the case of a memoryless system, the impulse response can be written as

$$h(t) = K\delta(t) \tag{2.24}$$

Recall that $\delta(t)$ is zero except at the origin. This allows us to determine whether or not a system has memory. So the converse to (2.24) is that a system has memory if $h(t) \neq 0$ when $t \neq 0$.

Causality in LTI Systems

When an LTI system is causal, the impulse response satisfies

$$h(t) = 0 \quad \text{when } t < 0 \tag{2.25}$$

Looking at (2.20), we see that for a casual system we can write the output signal as

$$y(t) = \int_0^\infty h(\tau)x(t - \tau)\,d\tau \tag{2.26}$$

A *causal signal* is one for which $x(t) = 0$ for $t < 0$, while an *anticausal signal* is one for which $x(t) = 0$ for $t > 0$.

Stability in LTI Systems

An LTI system is stable if

$$\int_{-\infty}^{\infty} |h(\tau)| \, d\tau < \infty \qquad (2.27)$$

EXAMPLE 2-4
An LTI system has the impulse response function

$$h(t) = e^{-t}u(t)$$

Is the system memoryless? Is it causal? Is it stable?

SOLUTION 2-4
We show a plot of this function in Fig. 2-13. The system is not memoryless, since $h(t)$ is not zero when $t \neq 0$.
   Since $h(t) = 0$ for $t < 0$, the system is causal.
   To determine if the system is stable, we integrate $|h(t)|$. We have

$$\int_{-\infty}^{\infty} |h(t)| \, dt = \int_{-\infty}^{\infty} e^{-t}u(t) \, dt = \int_{0}^{\infty} e^{-t} \, dt = -e^{-t} \Big|_{0}^{\infty} = 1$$

Since this impulse response satisfies $\int_{-\infty}^{\infty} |h(\tau)| \, d\tau < \infty$, the system is stable.

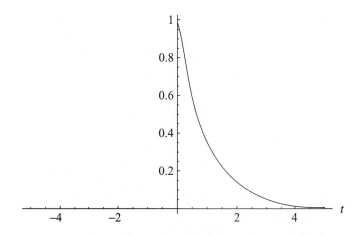

Fig. 2-13. The impulse response function for Example 2-4.

EXAMPLE 2-5
Consider an impulse response given by

$$h(t) = 4\delta(t)$$

Discuss the properties of the system.

SOLUTION 2-5
The first item to notice is that $h(t) = 0$ when $t \neq 0$, and therefore the system is memoryless. It immediately follows that the system is causal. To determine stability, notice that

$$\int_{-\infty}^{\infty} |h(\tau)| d\tau = \int_{-\infty}^{\infty} 4\delta(t) \, dt = 4 \int_{-\infty}^{\infty} \delta(t) \, dt = 4 < \infty$$

Therefore the system is stable.

EXAMPLE 2-6
Discuss the memory, causality, and stability of the system with impulse response

$$h(t) = \cos(\pi t)u(t + 1)$$

SOLUTION 2-6
It might be helpful to plot the function. This plot is shown in Fig. 2-14.

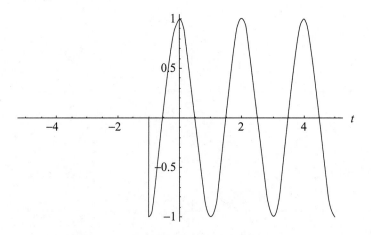

Fig. 2-14. A plot of $h(t) = \cos(\pi t)u(t + 1)$.

Since $h(t) \neq 0$ when $t \neq 0$, the system is not memoryless. Moreover, notice that the impulse response is nonzero for values of $t < 0$. This tells us that the system is not causal. To determine stability, we consider

$$\int_{-\infty}^{\infty} |h(\tau)| d\tau = \int_{-\infty}^{\infty} \cos(\pi t) u(t+1) \, dt = \int_{-1}^{\infty} \cos(\pi t) \, dt = \infty$$

Since the integral of the impulse response function is not convergent, the system is not stable.

Convolution

We now turn to a more detailed study of the convolution of two functions, a very important operation to consider when looking at LTI systems. We write the convolution of two functions f and g as

$$f * g = \int_{-\infty}^{\infty} f(\tau) g(t - \tau) \, d\tau \tag{2.28}$$

The convolution operation

- is commutative, i.e., $f * g = g * f$
- is associative, i.e., $[f * g] * h = f * [g * h]$
- is distributive, i.e., $f * (g + h) = f * g + f * h$

EXAMPLE 2-7
Given $x(t) = u(t)$ and $h(t) = \cos(\pi t) u(t)$, find the response $y(t)$.

SOLUTION 2-7
Notice that $h(t)$ is zero when $t < 0$, so the system is causal. To find the response to the given input signal (which in this case is the step response) we apply (2.26), which applies in the case of a causal system. We obtain

$$y(t) = \int_0^{\infty} h(\tau) x(t - \tau) \, d\tau = \int_0^{\infty} \cos(\pi \tau) u(t - \tau) \, d\tau$$

To do the integral, we rely on a graphical approach to find the limits of integration. First we reflect $x(t - \tau) = u(t - \tau)$ about the origin. First recall that

$$u(t - \tau) = \begin{cases} 1 & t > \tau \\ 0 & t < \tau \end{cases}$$

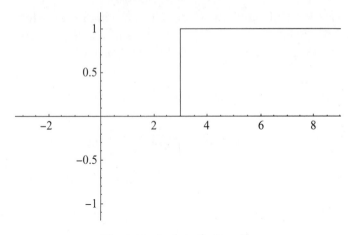

Fig. 2-15. A plot of $u(t - 3)$.

For some $\tau > 0$, the plot of $x(t - \tau) = u(t - \tau)$ looks like that shown in Fig. 2-15.

If we reflect it about the y-axis and shift, we obtain the function shown in Fig. 2-16.

The leading edge of the reflected and shifted $x(t - \tau)$ is at the value t, as indicated in the figure. Now we imagine sliding the function along and multiplying it by $h(\tau)$. Since $h(\tau)$ is only nonzero for positive values of τ, the functions do not overlap when $t < 0$ and so there is no contribution to the integral and $y(t) = 0$. When $t > 0$, the functions are both nonzero over the range $0 \leq \tau \leq t$. This is shown in Fig. 2-17.

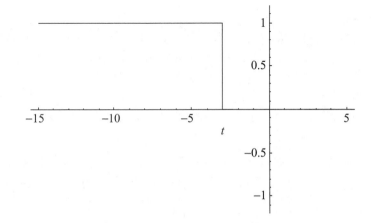

Fig. 2-16. The reflected and shifted function.

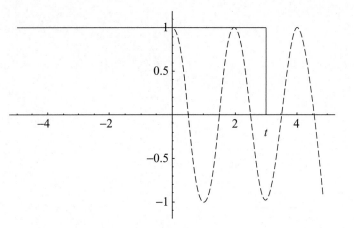

Fig. 2-17. The region of nonzero overlap is $0 \leq \tau \leq t$.

Therefore the system response is

$$y(t) = \int_0^t \cos(\pi \tau) \, d\tau = \frac{1}{\pi} \sin(\pi \tau) \bigg|_0^t = \frac{1}{\pi} \sin(\pi t)$$

EXAMPLE 2-8
Suppose that $x(t) = e^{-t}u(t+2)$ and $h(t) = e^{t}u(-t)$. Find the response $y(t)$.

SOLUTION 2-8
First let's plot the functions. $x(t) = e^{-t}u(t+2)$ is shown in Fig. 2-18.

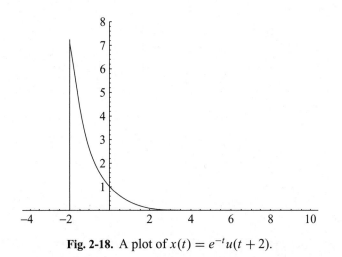

Fig. 2-18. A plot of $x(t) = e^{-t}u(t+2)$.

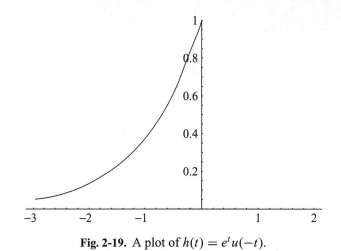

Fig. 2-19. A plot of $h(t) = e^t u(-t)$.

In Fig. 2-19, we have a plot of $h(t) = e^t u(-t)$.
To find the response of the system, we use

$$y(t) = \int_{-\infty}^{\infty} x(\tau) h(t - \tau) \, d\tau$$

Since $x(t) = e^{-t} u(t + 2)$, we need to calculate the convolution for two regions. These are $t < -2$ and $t \geq -2$. First we consider $t < -2$. We reflect h about the origin and shift it to the left. The trailing edge of h indicates the location of t.

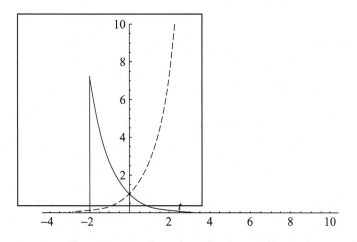

Fig. 2-20. The nonzero region of overlap ranges from -2 to ∞.

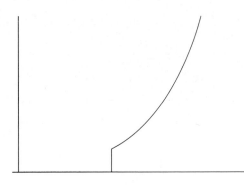

Fig. 2-21. A plot of $h(t - \tau)$ for $t \geq -2$, with t given by the trailing edge of the function.

We show both functions on the same plot in Fig. 2-20. The solid line is $x(\tau)$ and the dashed line is $h(t - \tau)$.

The nonzero region of overlap is found to be for τ between -2 and ∞, so we have

$$y(t) = \int_{-2}^{\infty} e^{-\tau} e^{t-\tau} d\tau = e^t \int_{-2}^{\infty} e^{-2\tau} d\tau = -\frac{e^t}{2} e^{-2\tau} \Big|_{-2}^{\infty} = \frac{e^{4+t}}{2}$$

Now, when $t \geq -2$, we have $h(t - \tau)$ as shown in Fig. 2-21 for some value of t.

In this case, we integrate for all $\tau \geq t$ and so we have

$$y(t) = \int_{t}^{\infty} e^{-\tau} e^{t-\tau} d\tau = e^t \int_{t}^{\infty} e^{-2\tau} d\tau = -\frac{e^t}{2} e^{-2\tau} \Big|_{t}^{\infty} = \frac{e^{-t}}{2}$$

Quiz

1. Let $y(t) = tx(t)$. Is the system memoryless? Is it time-invariant?
2. Is the system $y(t) = 1/x(t)$ linear?
3. Is the system $y(t) = x(8 - t)$ time-invariant?
4. Is the system $y(t) = t^2(dx/dt)$ linear?
5. Is the system $y(t) = t^2(dx/dt)$ stable?
6. Is the system $y(t) = t^2(dx/dt)$ memoryless? Is it time-invariant?

7. Consider the impulse response $h(t) = 5\delta(t)$. Is this system memoryless, casual, and stable? What is $\int_{-\infty}^{\infty} |h(\tau)| \, d\tau$?

8. Compute $\int_0^\infty e^{-t} \cos(\pi t) \delta(t - 2) \, dt$.

9. Find the step response for $h(t) = e^{-5t} u(t)$.

10. Find $\int_{-\infty}^{\infty} t^2 \delta(t + 5) \, dt$.

CHAPTER 3

Discrete Time Signals

In this chapter we carry over the results developed so far to the case of *discrete time signals*. To review, a discrete time signal is one for which the time variable is defined only at specific discrete times. Such a signal can be realized by sampling a continuous time signal $x(t)$ to form a collection or list of numbers $x[n]$. If we denote each time step that we sample as t_i, then the list of numbers that make up the discrete time signal are calculated by evaluating the continuous time signal $x(t)$ at each time step, i.e., we compute $x(t_i)$. We can calculate the $x(t_i)$ for several values and arrange them in a list, which we display here schematically

$$x(t_0), x(t_1), x(t_2), \ldots, x(t_n), \ldots \tag{3.1}$$

The difference between successive time steps is called the *sampling interval*. Often, the sampling interval is uniform and each time step can be labeled by an integer n. So we can write

$$x[0], x[1], x[2], \ldots, x[n], \ldots \tag{3.2}$$

If the sampling interval (which is a certain length of time) is denoted by T_s, then the discrete time signal is a sequence of numbers given by

$$x[n] = x(nT_s) \tag{3.3}$$

where $x(nT_s)$ is some continuous time signal evaluated or sampled at the time $t = nT_s$. It is common to denote a sequence of numbers by $\{x_n\}$. If we have two discrete time signals $\{x_n\}$ and $\{y_n\}$, we can form the sum and product of these signals by adding them together term by term in the usual way, that is

$$\{z_n\} = \{x_n\} + \{y_n\} \Rightarrow z[n] = x[n] + y[n] \tag{3.4}$$

$$\{z_n\} = \{x_n\}\{y_n\} \Rightarrow z[n] = x[n]y[n] \tag{3.5}$$

We can also construct a new sequence by scalar multiplication. Given a constant α,

$$\{y_n\} = \alpha\{x_n\} \Rightarrow y[n] = \alpha x[n] \tag{3.6}$$

To plot a sequence, we simply put a point at the appropriate location and draw a line up from the time axis to the point. This is best illustrated by example. Suppose that we have a discrete time signal constructed by sampling $\cos t$ at a uniform interval of $\pi/6$ over the range $0 \leq t \leq 2\pi$. The time samples are shown in Table 3-1.

We plot these values in Fig. 3-1.

EXAMPLE 3-1
Consider two sequences

$$\{x_n\} = \{-1, 3, 2, 4, -2, 1\}$$
$$\{y_n\} = \{0, 1, 1, -2, -2, 2\}$$

Plot each sequence assuming that $n = 1$ is the first term. Then construct their sum and product, and construct new sequences using scalar multiplication where $\alpha = 3$.

SOLUTION 3-1
The plot of $\{x_n\}$ is shown in Fig. 3-2.

The plot of $\{y_n\}$ is shown in Fig. 3-3.

Table 3-1. A discrete time signal constructed by sampling the cosine function with a sampling interval of $T_s = \pi/6$.

Index (n)	Time (nT_s)	$x[n] = \cos(nT_s)$
0	0	1
1	0.524	0.866
2	1.047	0.5
3	1.571	0
4	2.094	−0.5
5	2.618	−0.866
6	3.142	−1
7	3.665	0.866
8	4.189	−0.5
9	4.712	0
10	5.236	0.5
11	5.760	0.866
12	6.283	1

The sum is given by

$$\{z_n\} = \{x_n\} + \{y_n\} = \{-1, 3, 2, 4, -2, 1\} + \{0, 1, 1, -2, -2, 2\}$$
$$= \{-1 + 0, 3 + 1, 2 + 1, 4 - 2, -2 - 2, 1 + 2\}$$
$$= \{-1, 4, 3, 2, -4, 3\}$$

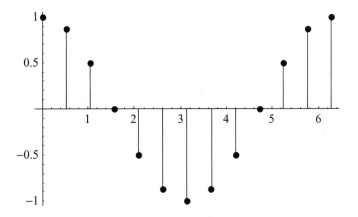

Fig. 3-1. A plot of the discrete time signal constructed in Table 3-1.

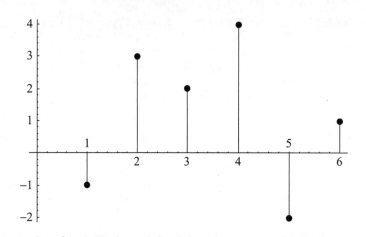

Fig. 3-2. A plot of $\{x_n\}$. We have defined these sequences with the first term given by $n = 1$.

The product is found by multiplying the individual terms

$$\{z_n\} = \{x_n\}\{y_n\} = \{-1, 3, 2, 4, -2, 1\}\{0, 1, 1, -2, -2, 2\}$$
$$= \{-1{*}0, 3{*}1, 2{*}1, 4{*}{-2}, -2{*}{-2}, 1{*}2\}$$
$$= \{0, 3, 2, -8, 4, 2\}$$

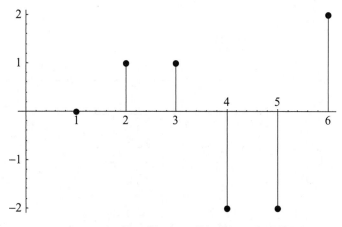

Fig. 3-3. A plot of $\{y_n\}$ used in Example 3-1.

For scalar multiplication by 3, we obtain

$$3\{x_n\} = 3\{-1, 3, 2, 4, -2, 1\} = \{-3, 9, 6, 12, -6, 3\}$$
$$3\{y_n\} = 3\{0, 1, 1, -2, -2, 2\} = \{0, 3, 3, -6, -6, 6\}$$

Digital Signals

A discrete time signal is called a *digital signal* if $x[n]$ can assume only a finite number of distinct values.

Energy and Power in Discrete Signals

The normalized energy content of a discrete time signal is given by

$$E = \sum_{-\infty}^{\infty} |x[n]|^2 \tag{3.7}$$

while the average power is

$$P = \lim_{N \to \infty} \frac{1}{2N+1} \sum_{n=-N}^{N} |x[n]|^2 \tag{3.8}$$

Previous classification results of signals as energy or power signals carry over from the continuous case. In particular,

- if $0 < E < \infty$ and $P = 0$ then we say that $x[n]$ is an energy signal;
- if $0 < P < \infty$ and $E = \infty$ then $x[n]$ is a power signal;
- if neither condition is satisfied, then $x[n]$ is neither an energy signal nor a power signal.

EXAMPLE 3-2
Suppose that

$$\{x_n\} = \sum_{n=0}^{\infty} \left(\frac{1}{2}\right)^n$$

Find the energy content of this signal. Is this an energy signal or a power signal?

SOLUTION 3-2
Squaring each term, we have

$$|x[n]|^2 = \left|\left(\frac{1}{2}\right)^n\right|^2 = \left(\frac{1}{4}\right)^n$$

And so we have

$$E = \sum_{n=0}^{\infty} \left(\frac{1}{4}\right)^n$$

This is nothing more than a geometric series. If $|r| < 1$, then

$$\sum_{k=0}^{\infty} ar^k = \frac{a}{1-r}$$

This is certainly true in this case, where $r = 1/4$ and $a = 1$ and so

$$E = \sum_{n=0}^{\infty} \left(\frac{1}{4}\right)^n = \frac{1}{1-(1/4)} = \frac{1}{3/4} = \frac{4}{3}$$

The energy content satisfies $0 < E < \infty$. For the average power, we have

$$P = \lim_{N \to \infty} \frac{1}{2N+1} \sum_{n=0}^{N} |x[n]|^2$$

(the series starts at 0 since by definition the signal vanishes before then). Now we already know that in the limit $\sum_{n=-N}^{N} |x[n]|^2$ goes to 4/3. So we only need to worry about

$$P = \lim_{N \to \infty} \frac{1}{2N+1} = 0$$

So, we conclude that the signal is an energy signal.

EXAMPLE 3-3
Suppose that

$$\{x_n\} = \sum_{n=1}^{\infty} \frac{1}{n}$$

Find the energy content of the signal.

SOLUTION 3-3

Squaring each term we find

$$E = \sum_{n=1}^{\infty} \frac{1}{n^2}$$

This series is convergent; in fact

$$E = \sum_{n=1}^{\infty} \frac{1}{n^2} = \frac{\pi^2}{6}$$

The Unit Impulse Sequence

In the next two sections we define discrete signals that are analogous to the unit step and unit impulse functions examined in the previous chapter. We begin by defining the *unit impulse sequence*. This sequence is denoted by $\delta[n]$ and has the following rather simple definition

$$\delta[n] = \begin{cases} 1 & n = 0 \\ 0 & n \neq 0 \end{cases} \tag{3.9}$$

In further analogy to the unit impulse function, we can shift the unit impulse sequence by some value k, where k is an integer to give

$$\delta[n - k] = \begin{cases} 1 & n = k \\ 0 & n \neq k \end{cases} \tag{3.10}$$

In Fig. 3-4, we show $\delta[n - 3]$.

It should be clear that this sequence also satisfies the sampling property. That is

$$x[n]\delta[n - k] = x[k]\delta[n - k] \tag{3.11}$$

For example, consider the discrete time signal $x[n]$ constructed by sampling $\cos t$ illustrated in Table 3-1. We have

$$x[n]\delta[n - 5] = x[5]\delta[n - 5] = (-0.866)\delta[n - 5] = (-0.866)(1) = -0.866$$

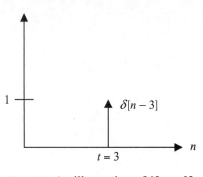

Fig. 3-4. An illustration of $\delta[n-3]$.

Moreover, we can express any sequence $x[n]$ in terms of the unit impulse sequence as follows:

$$x[n] = \sum_{k=-\infty}^{\infty} x[k]\delta[n-k] \qquad (3.12)$$

The Unit Step Sequence

The unit step sequence is defined in the following way:

$$u[n] = \begin{cases} 1 & n \geq 0 \\ 0 & n < 0 \end{cases}$$

This is just a sequence of 1s starting at the origin. We illustrate this in Fig. 3-5. We can shift the sequence to define

$$u[n-k] = \begin{cases} 1 & n \geq k \\ 0 & n < k \end{cases} \qquad (3.13)$$

This is illustrated in Fig. 3-6.

By combining different unit step sequences together, we can construct a "square pulse" when $j < k$ using $u[n-j] - u[n-k]$. Specifically, consider $u[n-2] - u[n-5]$. These are shown individually in Fig. 3-7.

It is clear that their difference gives the sequence shown in Fig. 3-8.

Notice that we can define the unit impulse sequence in terms of the unit step sequence as follows:

$$\delta[n] = u[n] - u[n-1] \qquad (3.14)$$

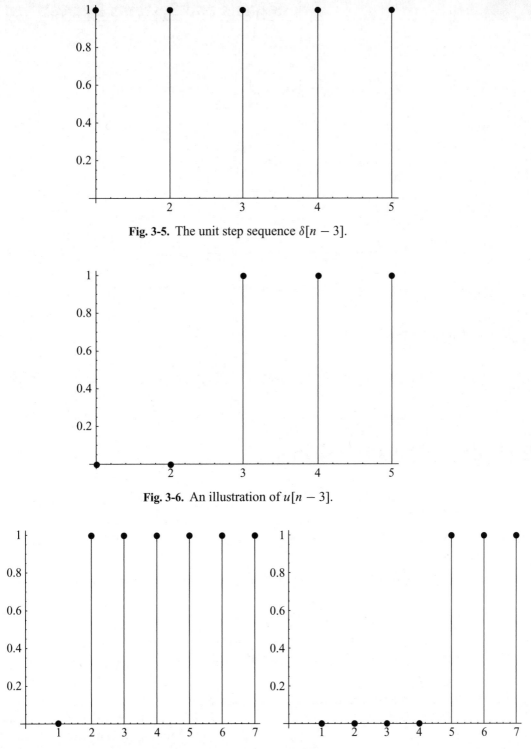

Fig. 3-5. The unit step sequence $\delta[n-3]$.

Fig. 3-6. An illustration of $u[n-3]$.

Fig. 3-7. An illustration of $u[n-2]$ and $u[n-5]$.

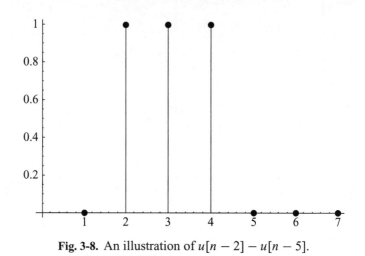

Fig. 3-8. An illustration of $u[n-2] - u[n-5]$.

Conversely, we have

$$u[n] = \sum_{k=-\infty}^{n} \delta[k] \qquad (3.15)$$

Periodic Discrete Signals

A discrete time signal $n[x]$ is called *periodic* if there exists a positive integer N such that

$$x[n+N] = x[n] \, \forall n \qquad (3.16)$$

We call the smallest N for which (3.16) holds the *fundamental period* and denote it by N_0.

EXAMPLE 3-4
Consider the discrete time signal constructed by sampling a sawtooth wave shown in Fig. 1-4. What is the fundamental period?

SOLUTION 3-4
For convenience we reproduce the sequence in Fig. 3-9.

Starting at the origin where $x[0] = 0$, notice that the next time the signal attains the value 0 at $x[2] = 0$. We can see this going in the negative direction

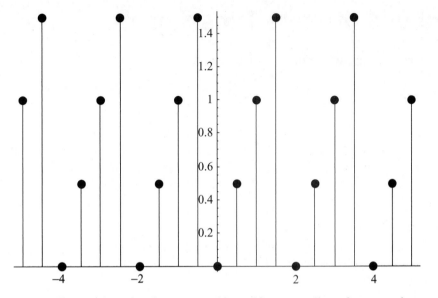

Fig. 3-9. A discrete time signal constructed by uniform sampling of a sawtooth wave.

as well, so we have

$$x[-4] = x[-2] = x[0] = x[2] = 0 \qquad (3.17)$$

Therefore, we conclude that $N_0 = 2$ is the fundamental period and that this sequence satisfies $x[n + 2] = x[n] \, \forall n$.

The *fundamental frequency* of a discrete time signal is defined using

$$\omega_0 = \frac{2\pi}{N_0} \text{ rad/sample} \qquad (3.18)$$

EXAMPLE 3-5
What is the fundamental frequency for the discrete time signal used in Example 3-4?

SOLUTION 3-5
We found that $N_0 = 2$, and therefore

$$\omega_0 = 2\pi/N_0 = 2\pi/2 = \pi \text{ rad/sample}$$

Even and Odd Discrete Time Signals

A discrete time signal $x[n]$ is called *even* if

$$x[-n] = x[n] \tag{3.19}$$

A discrete time signal $x[n]$ is called *odd* if

$$x[-n] = -x[n] \tag{3.20}$$

Figure 3-10 shows an example of an even discrete time signal.

In Fig. 3-11, we show an odd discrete time signal.

Given an arbitrary discrete time signal $x[n]$, we can construct the even and odd parts using

$$x_e[n] = \frac{x[n] + x[-n]}{2}, \qquad x_o[n] = \frac{x[n] - x[-n]}{2} \tag{3.21}$$

EXAMPLE 3-6

Consider the discrete time signal shown in Fig. 3-12.
  (a) Write down the members of the sequence.
  (b) Sketch $x[n]u[n]$.
  (c) Sketch $x[n]u[n-2]$.
  (d) Sketch $x[n]\delta[n-1]$.

SOLUTION 3-6
  (a) The sequence is $\{1, 2, 3, 0, 4, 1, 3\}$.

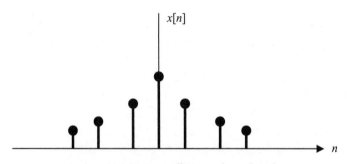

Fig. 3-10. An even discrete time signal.

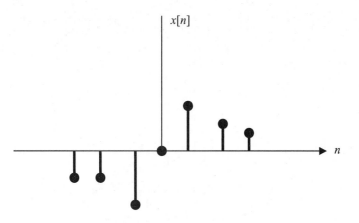

Fig. 3-11. An odd discrete time signal.

(b) To sketch $x[n]u[n]$, first recall that

$$u[n - k] = \begin{cases} 1 & n \geq k \\ 0 & n < k \end{cases}$$

With $k = 0$, all members of $u[n]$ with $n < 0$ are zero. To find the members of the product, we use (3.5). We obtain $\{0, 4, 1, 3\}$. This is shown in Fig. 3-13.

(c) To find $x[n]u[n - 2]$, now we note that the only terms that survive will be those with $n \geq 2$. We have $x[n]u[n - 2] = \{1, 3\}$. This is shown in Fig. 3-14.

(d) Finally, we note that $\delta[n - 1] = 1$ for $n = 1$ and is zero otherwise. The sampling property tells us that $x[n]\delta[n - 1] = x[1]$, so we obtain the sketch in Fig. 3-15.

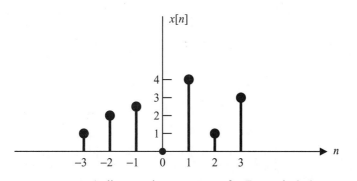

Fig. 3-12. A discrete time sequence for Example 3-6.

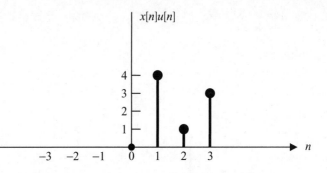

Fig. 3-13. The product $x[n]u[n]$ for Example 3-6.

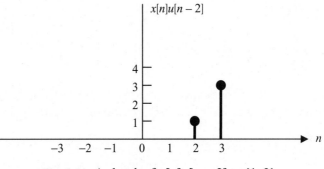

Fig. 3-14. A sketch of $x[n]u[n-2] = \{1, 3\}$.

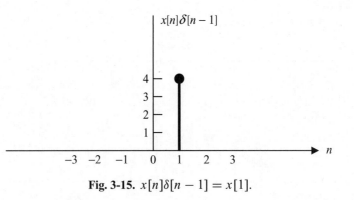

Fig. 3-15. $x[n]\delta[n-1] = x[1]$.

Properties of Discrete Time Signals

We now examine linearity, memory, causality, stability, and time-invariance for discrete time signals. The definitions follow from those used with continuous time signals. To check linearity, we need to check to see if $\hat{T}\{ax_1[n] + bx_2[n]\} = a\hat{T}\{x_1[n]\} + b\hat{T}\{x_2[n]\}$ is satisfied.

EXAMPLE 3-7
Determine whether or not $y[n] = x[n - 2]$ and $y[n] = x[n]u[n - 1]$ are linear.

SOLUTION 3-7
Considering $y[n] = x[n - 2]$ first, we have

$$\hat{T}\{ax_1[n] + bx_2[n]\} = ax_1[n - 2] + bx_2[n - 2] = ay_1[n] + by_2[n]$$

Therefore $y[n] = x[n - 2]$ is linear. For the second case, we have

$$\hat{T}\{ax_1[n] + bx_2[n]\} = (ax_1[n] + bx_2[n])u[n - 1]$$
$$= ax_1[n]u[n - 1] + bx_2[n]u[n - 1]$$
$$= ay_1[n] + by_2[n]$$

The second transformation is also linear.

Following the definition used for continuous time signals, if a transformation does not depend on previous members of the sequence, it is *memoryless*. Put another way, if the output value at n depends on input values at previous values of n (such as $n - 1$ say), then the system is not memoryless.

EXAMPLE 3-8
Determine whether $y[n] = x[n - 2]$ and $y[n] = 2x[n]u[n]$ are memoryless.

SOLUTION 3-8
The system described by $y[n] = x[n - 2]$ is not memoryless because the output value at n depends on the input values at $n - 2$. However, the output value at n for $y[n] = 2x[n]u[n]$ depends only on input values at n, and therefore this system is memoryless.

If a system output value at n does not depend on future input values, then it is *causal*. Also recall that a memoryless system is automatically causal.

EXAMPLE 3-9
Determine whether or not the systems $y[n] = 3x[n - 1]$ and $y[n] = 2x^2[n + 1]$ are causal.

SOLUTION 3-9

Since the output value at n for the system described by $y[n] = 3x[n-1]$ depends only on previous values of n, the system is causal. For $y[n] = 2x^2[n+1]$, the output value at n depends on the input value at $n+1$, so it is not causal.

Next we recall stability. If $|x[n]| \leq k \; \forall n$ implies that $|y[n]| \leq k$, then the system is bounded-input, bounded-output (BIBO) stable.

EXAMPLE 3-10

Are the systems $y[n] = 3x[n-1]$ and $y[n] = 2xn[n]$ stable?

SOLUTION 3-10

Assume that $x[n] \leq k$ for some finite k and for all n. In the first case $x[n] \leq k$ implies that $|y[n]| \leq 3k$ and so the system is stable. In the second case, since we have $y[n] = 2nx[n]$ where the output depends on n directly, it grows without bound as n increases, so it is not stable.

Finally, we review time-invariance. A discrete time system is time-invariant if the following holds:

$$\hat{T}\{x[n-k]\} = y[n-k] \tag{3.22}$$

EXAMPLE 3-11

Determine whether or not $y[n] = 3x[n-1]$ and $y[n] = 2nx[n]$ are time-invariant.

SOLUTION 3-11

Given $y[n] = 3x[n-1]$, we see that $y[n-k] = 3x[n-1-k]$. Now we have

$$\hat{T}\{x[n-k]\} = 3x[n-k-1] = 3x[n-1-k] = y[n-k]$$

So, the system is time-invariant. For the second case, given $y[n] = 2nx[n]$ we have $y[n-k] = 2(n-k)x[n-k]$. The transformation applied to $x[n-k]$ gives

$$\hat{T}\{x[n-k]\} = 2nx[n-k] \neq y[n-k]$$

So this system is *not* time-invariant.

Discrete Linear Time-Invariant Systems

The definitions in this section follow from those in Chapter 2 involving continuous time systems. We begin by defining the impulse response, which in this case is the response of a system to the unit impulse function

$$h[n] = \hat{T}\{\delta[n]\}$$ (3.23)

The system is time-invariant, so we can write $h[n-k] = \hat{T}\{\delta[n-k]\}$. Putting this together with (3.12), we see that the response to an arbitrary input $x[n]$ is given by

$$y[n] = \sum_{k=-\infty}^{\infty} x[k]h[n-k]$$ (3.24)

The right-hand side of (3.24) is a discrete version of convolution. That is, we have

$$y[n] = x[n]^*h[n]$$

The commutative, associative, and distributive properties also apply in the discrete case. Since convolution is commutative, we can write

$$y[n] = h[n]^*x[n] = \sum_{k=-\infty}^{\infty} h[k]x[n-k]$$ (3.25)

Convolution is best described using examples.

EXAMPLE 3-12
Compute $y[n] = x[n]^*h[n]$ for the sequences shown in Figs. 3-16 and 3-17.

SOLUTION 3-12
We start with a small example to illustrate a graphical shift and add algorithm that can be used to compute convolution in the discrete case. What we do is compute new sequences that are based on shifting and scaling of $x[n]$ based on

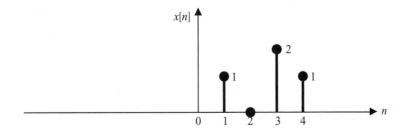

Fig. 3-16. An input signal $x[n]$.

Signals and Systems Demystified

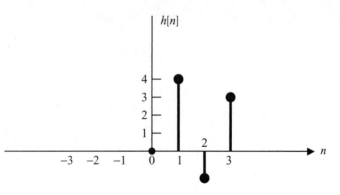

Fig. 3-17. The impulse response for Example 3-12.

the data contained in $h[n]$, and then add these sequences together to obtain $y[n]$. Looking at the data, $h[n]$ has three nonzero data points, so we need to create three shifted and scaled copies of $x[n]$. The first nonzero data point occurs at $n = +1$ and we have $h[1] = 4$. This tells us that we need to compute $4x[n - 1]$. We can do this by shifting each data point of $x[n]$ one position to the right, and then we multiply by 4. The result is shown in Fig. 3-18.

The next data point for $h[n]$ is located at $n = 2$. This tells us that the next data set will be shifted as $x[n - 2]$. Since $h[2] = -1$, the sequence we need is $-x[n - 2]$, which we show in Fig. 3-19.

The final data point in $h[n]$ is $h[3] = 3$. This tells us the next sequence to compute is $3x[n - 3]$. This is shown in Fig. 3-20.

To obtain $y[n]$, we simply add up all the terms. We obtain

$$y[0] = 0$$
$$y[1] = 0$$
$$y[2] = 4 + 0 + 0 = 4$$
$$y[3] = 0 - 1 + 0 = -1$$

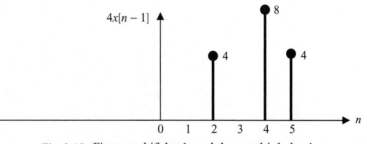

Fig. 3-18. First we shift by 1, and then multiply by 4.

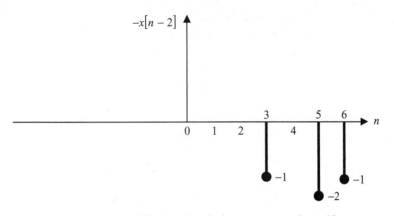

Fig. 3-19. The shifted and scaled sequence $-x[n-2]$.

$$y[4] = 8 + 0 + 3 = 11$$
$$y[5] = 4 - 2 + 0 = 2$$
$$y[6] = 0 - 1 + 6 = 5$$
$$y[7] = 0 + 0 + 3 = 3$$

The resulting sequence is shown in Fig. 3-21.

EXAMPLE 3-13
Find $y[n] = x[n]*h[n]$ for the signals shown in Figs. 3-22 and 3-23.

SOLUTION 3-13
Figure 3-22 shows $x[n]$ while Fig. 3-23 shows $h[n]$. Looking at $h[n]$, we see that we can find $y[n] = x[n]*h[n]$ if we have $x[n+1], x[n], 2x[n-2]$.

Using the procedure in the previous example, we plot $x[n+1], x[n]$, $2x[n-2]$ and add up the terms. We already have $x[n]$; we can obtain $x[n+1]$ by shifting this one unit to the left. This is shown in Fig. 3-24.

In Fig. 3-25, we show $2x[n-2]$.

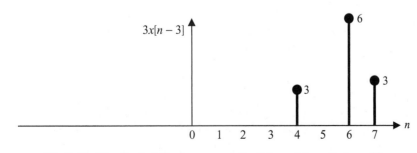

Fig. 3-20. The final shifted sequence for this problem, $3x[n-3]$.

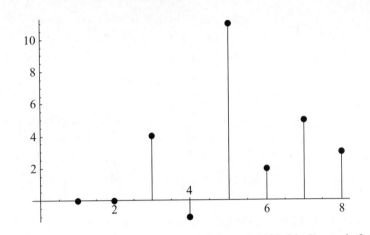

Fig. 3-21. The sequence obtained from, $y[n] = x[n]*h[n]$ in Example 3-12.

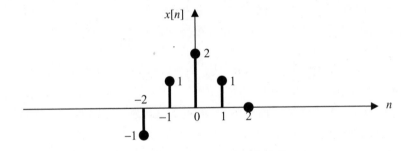

Fig. 3-22. The input signal $x[n]$.

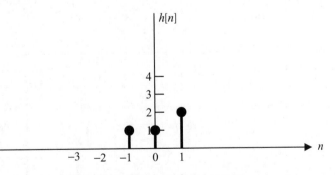

Fig. 3-23. The impulse response.

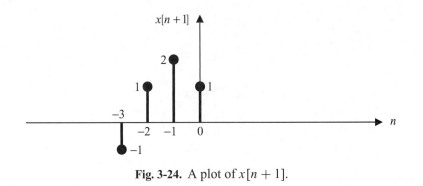

Fig. 3-24. A plot of $x[n + 1]$.

Looking at the plots in Figs. 3-22, 3-24, and 3-25, we add the corresponding terms to get the output signal. We find

$$y[-3] = -1$$
$$y[-2] = -1 + 1 = 0$$
$$y[-1] = 1 + 2 = 3$$
$$y[0] = 2 + 1 - 2 = 1$$
$$y[1] = 1 + 2 = 3$$
$$y[2] = 4$$
$$y[3] = 2$$

The output signal is shown in Fig. 3-26.

A general convolution algorithm to find $y[n]$ is to follow these steps (note that n is some fixed value; it will range over all values for which $y[n]$ is not zero):

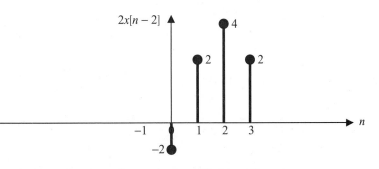

Fig. 3-25. A plot of $2x[n - 2]$.

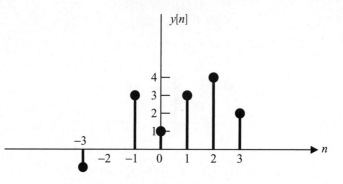

Fig. 3-26. The output signal for Example 3-13.

- Flip $h[k]$ about the origin to give $h[-k]$
- Shift $h[-k]$ to give $h[-(k-n)]$
- Multiply the terms of $x[k]$ and $h[-(k-n)]$ together and sum
- Repeat until there is no overlap

Memoryless Systems

A memoryless discrete linear time-invariant (LTI) system has the form

$$y[n] = Kx[n] \tag{3.26}$$

where K is a gain constant. For the impulse response, this takes the form

$$h[n] = K\delta[n] \tag{3.27}$$

A system is memoryless if $h[n] = 0$ when $n \neq 0$, otherwise the system has memory.

CAUSALITY

A memoryless system is causal. More generally, a discrete LTI system is causal if

$$h[n] = 0 \tag{3.28}$$

for $n < 0$.

STABILITY

A discrete time LTI system is BIBO stable if

$$\sum_{k=-\infty}^{\infty} |h[k]| < \infty \qquad (3.29)$$

EXAMPLE 3-14

Suppose that the impulse response for a discrete LTI system is

$$h[n] = nu[n]$$

Is the system memoryless? Is it causal? Is it BIBO stable?

SOLUTION 3-14

Since there are terms for which $h[n] \neq 0$ when $n \neq 0$, the system is not memoryless. However, since $h[n] = 0$ for $n < 0$, the system is causal. To investigate stability, we compute the sum of $|h[n]|$. We have

$$\sum_{k=-\infty}^{\infty} |h[k]| = \sum_{k=-\infty}^{\infty} |ku[k]| = \sum_{k=0}^{\infty} k \to \infty$$

Since $\sum_{k=-\infty}^{\infty} |h[k]|$ is not finite, the system is not BIBO stable.

EXAMPLE 3-15

Suppose that the impulse response for a discrete LTI system is

$$h[n] = \alpha^{n-1} u[n-2]$$

where $|\alpha| < 1$. Is the system memoryless? Is it causal? Is it BIBO stable?

SOLUTION 3-15

The system is not memoryless, because there are terms for which $h[n] \neq 0$ when $n \neq 0$. The system is causal because $h[n] = 0$ for $n < 0$. Now,

$$\sum_{k=-\infty}^{\infty} |h[k]| = \sum_{k=-\infty}^{\infty} |\alpha^{k-1} u[k-2]| = \sum_{k=2}^{\infty} \alpha^{k-1}$$

We can rewrite this series in a more convenient form with a change of variable. We assume $m = k - 1$. When $k = 2$ we have $m = 1$ and the series becomes

$$\sum_{k=2}^{\infty} |\alpha|^{k-1} = \sum_{m=1}^{\infty} |\alpha|^m = \frac{|\alpha|}{|\alpha| - 1}$$

This means that $\sum_{k=-\infty}^{\infty} |h[k]| < \infty$ and so the system is BIBO stable.

STEP RESPONSE

The step response for a discrete LTI is given by

$$s[n] = h[n]^*u[n] \tag{3.30}$$

EXAMPLE 3-16
Find the step response for a system with impulse response $h[n] = 3\delta[n] + \delta[n - 1]$.

SOLUTION 3-16
Using (3.30) and the definition of convolution given in (3.25), we have

$$s[n] = h[n]^*u[n] = \sum_{k=-\infty}^{\infty} h[n - k]u[k]$$

(remember, convolution is commutative). Now we use the fact that $u[k] = 0$ for $k < 0$ to write this as

$$s[n] = \sum_{k=0}^{\infty} h[n - k]u[k] = \sum_{k=0}^{\infty} h[n - k] = \sum_{k=0}^{\infty} 3\delta[n - k] + \delta[n - 1 - k]$$

Now we recall the relationship between the unit impulse and the unit step, given in (3.15). This allows us to write

$$\sum_{k=0}^{\infty} 3\delta[n - k] = 3\sum_{k=0}^{\infty} \delta[n - k] = 3u[n]$$

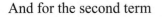

And for the second term

$$\sum_{k=0}^{\infty} \delta[n - 1 - k] = u[n - 1]$$

Therefore, the step response is

$$s[n] = 3u[n] + u[n - 1]$$

Quiz

1. A sequence $x[n]$ has a fundamental period given by $N_0 = 10$. What is the fundamental frequency?
2. What are the even and odd parts of a sequence?
3. If $x[n] = \sum_{n=0}^{\infty} \frac{1}{n}$ what is $x[n]\delta[n - 2]$?
4. Suppose that $\alpha = 1/3$ and a discrete time signal is given by $x[n] = \alpha^n u[n]$. What is the energy content of the signal?
5. Is $y[n] = nx[n]$ memoryless? Is it causal?
6. Is the system with impulse response $h[n] = n^2 u[n]$ causal?
7. Is the system with impulse response $h[n] = n^2 u[n]$ memoryless?
8. If $x[n] = \{1, 2, 1\}$ and $h[n] = \{1, 1, 1\}$ where $0 \le n \le 2$, find $y = x^*h$.
9. Consider an LTI system. If $h[n] = 2\delta[n]$, is the system memoryless?
10. Find the step response for $h[n] = \delta[n] - \delta[n - 3]$.

CHAPTER 4

Fourier Analysis

So far we have examined signals as functions of time. We now turn to a different task, the representation of a signal in the frequency domain. The Fourier series and Fourier transform provide two mathematical tools that allow us to transform a function of time into a function of frequency. When working with a function of time, we say that we have a time-domain signal. And when working with a function of frequency, we say that we have a frequency-domain signal. The frequency-domain representation of a signal is often called the *spectrum* of the signal. Both of these are two representations of the same object. As a loose analogy, think of a vector in the plane. We can represent the vector in Cartesian coordinates:

$$\vec{A} = A_x\hat{x} + A_y\hat{y}$$

Here the "components" of the vector, A_x and A_y, are numbers that describe the vector relative to Cartesian coordinates. The *same* vector can also be written in polar coordinates. In this case we have

$$\vec{A} = A_r\hat{r} + A_\theta\hat{\theta}$$

The components of the vector with respect to polar coordinates, A_r and A_θ, are numbers that are in general different than A_x and A_y, but this is the same vector, the same physical object lying in the plane. You can use your knowledge of vectors to gain a bit of understanding as to what we are doing with Fourier series and Fourier transforms. We have some signal which is a function of time, say $x(t)$. In the following sections we will learn how to obtain the Fourier transform of the signal, which is a function of frequency f. We write this as $X(f)$. Just as two expansions of a vector in two different coordinate systems are two representations of the same vector, the Fourier transform is the *same* signal represented with respect to frequency; that is, one signal, two representations.

Fourier Series

We begin our exploration of Fourier analysis by considering the representation of a periodic signal by a trigonometric series. Representations of this type originally came about using separation of variables techniques to solve partial differential equations. We aren't going to be worried about those kinds of details in this book; the interested reader can examine a text on partial differential equations. For our purposes we will need to only state what the Fourier series representation of a periodic signal is. First let's quickly review the definition of a periodic signal.

We say that a signal $x(t)$ is periodic if there is some $T > 0$, such that

$$x(t + T) = x(t) \tag{4.1}$$

for all t. T is called the *period* of the signal $x(t)$. The relation described in (4.1) tells us that the properties of the signal repeat themselves at regular intervals. Any periodic signal has many periods. We can find a new period by simply multiplying T by some integer m; that is,

$$x(t) = x(t + T) = x(t + 2T) = x(t + 3T) = \cdots = x(t + mT) = \cdots$$

The period of most interest to us is the one with the smallest value of T for which (4.1) holds. We call this the *fundamental period*, which is denoted by T_0. Using this definition, we can also define the fundamental frequency as

$$f_0 = \frac{1}{T_0} \tag{4.2}$$

which is measured in hertz (Hz). The fundamental angular frequency, which is less important in signal analysis, is given by $\omega_0 = 2\pi f_0$.

A periodic signal $x(t)$ can be represented by a *Fourier series expansion*. Such an expansion is given by

$$x(t) = a_0 + 2\sum_{n=1}^{\infty} \left[a_n \cos\left(\frac{2\pi nt}{T_0}\right) + b_n \sin\left(\frac{2\pi nt}{T_0}\right) \right] \qquad (4.3)$$

Notice the following: The trigonometric functions $\cos x$ and $\sin x$ have a fundamental period equal to 2π. Considering the first term in the series, where we will find the fundamental period of (4.3), the argument of each of the trigonometric terms is $2\pi t / T_0$. Now let's add the fundamental period of $\cos x$ and $\sin x$, 2π:

$$\frac{2\pi t}{T_0} + 2\pi = \frac{2\pi t}{T_0} + \frac{2\pi T_0}{T_0} = \frac{2\pi}{T_0}(t + T_0)$$

What we have learned from this little exercise is that both $\cos(2\pi nt / T_0)$ and $\sin(2\pi nt / T_0)$ are periodic in T_0, which is the fundamental period of the signal $x(t)$.

Given a periodic signal $x(t)$, the task at hand when constructing a Fourier series to obtain its frequency content is to find the expansion coefficients a_n and b_n. This is done by calculating a set of integrals that we will show in a minute. To see why you would do this, we go back to vectors. Let's write down our vector in Cartesian coordinates again:

$$\vec{A} = A_x \hat{x} + A_y \hat{y}$$

Which mathematical technique could you use to extract say the component A_x? You could do so by computing the dot product of $\vec{A}$ with the basis vector $\hat{x}$:

$$\hat{x} \cdot \vec{A} = \hat{x} \cdot (A_x \hat{x} + A_y \hat{y}) = A_x \hat{x} \cdot \hat{x} + A_y \hat{x} \cdot \hat{y} = A_x$$

We are able to do this since $\hat{x} \cdot \hat{x} = 1$ and $\hat{x} \cdot \hat{y} = 0$. Mathematically, the sort of expansion we are doing with a Fourier series is the same kind of animal. In this case $(2\pi nt / T_0)$ and $\sin(2\pi nt / T_0)$ play the same kind of role as the basis vectors $\hat{x}$ and $\hat{y}$ do for vectors. For this reason, we call the cosine and sine functions used in (4.3) *basis functions*. They obey a set of relations that are analogous to saying that $\hat{x} \cdot \hat{x} = 1$ and $\hat{x} \cdot \hat{y} = 0$ for basis vectors, which we'll call *orthogonality relations*. These are

$$\int_{-T_0/2}^{T_0/2} \cos\left(\frac{2\pi mt}{T_0}\right) \sin\left(\frac{2\pi nt}{T_0}\right) dt = 0 \qquad (4.4)$$

$$\int_{-T_0/2}^{T_0/2} \cos\left(\frac{2\pi m t}{T_0}\right) \cos\left(\frac{2\pi n t}{T_0}\right) dt = \begin{cases} T_0/2 & \text{for } m = n \\ 0 & \text{for } m \neq n \end{cases} \qquad (4.5)$$

$$\int_{-T_0/2}^{T_0/2} \sin\left(\frac{2\pi m t}{T_0}\right) \sin\left(\frac{2\pi n t}{T_0}\right) dt = \begin{cases} T_0/2 & \text{for } m = n \\ 0 & \text{for } m \neq n \end{cases} \qquad (4.6)$$

Now considering what happens when we integrate each of the basis functions over a single period, we obtain

$$\int_{-T_0/2}^{T_0/2} \cos\left(\frac{2\pi n t}{T_0}\right) dt = \frac{T_0}{2\pi n} \sin\left(\frac{2\pi n t}{T_0}\right)\Bigg|_{-T_0/2}^{T_0/2} = \frac{T_0}{2\pi n}[2\sin(\pi n)] = 0$$

A similar result is obtained when integrating $\sin(2\pi n t / T_0)$. So if we integrate (4.3) over one period, we can obtain the leading coefficient of the expansion, a_0, which is the *mean value* of the signal over one period.

THE MEAN VALUE OF A PERIODIC SIGNAL

Integrating (4.3) over one period, we have

$$\int_{-T_0/2}^{T_0/2} x(t)\, dt = a_0 \int_{-T_0/2}^{T_0/2} dt + 2\int_{-T_0/2}^{T_0/2}\left\{\sum_{n=1}^{\infty}\left[a_n \cos\left(\frac{2\pi n t}{T_0}\right)\right.\right.$$
$$\left.\left. + b_n \sin\left(\frac{2\pi n t}{T_0}\right)\right]\right\} dt$$

We can bring the integral inside the sum on the right-hand side. From the discussion in the previous paragraph, we know that these terms vanish. That is

$$2\int_{-T_0/2}^{T_0/2}\left\{\sum_{n=1}^{\infty}\left[a_n \cos\left(\frac{2\pi n t}{T_0}\right) + b_n \sin\left(\frac{2\pi n t}{T_0}\right)\right]\right\} dt$$
$$= 2\left\{\sum_{n=1}^{\infty} a_n \int_{-T_0/2}^{T_0/2} \cos\left(\frac{2\pi n t}{T_0}\right) dt + b_n \int_{-T_0/2}^{T_0/2} \sin\left(\frac{2\pi n t}{T_0}\right) dt\right\} = 0$$

This leaves us with

$$\int_{-T_0/2}^{T_0/2} x(t)\,dt = a_0 \int_{-T_0/2}^{T_0/2} dt = a_0 t \Big|_{-T_0/2}^{T_0/2} = a_0 T_0$$

Solving for the coefficient a_0, we obtain

$$a_0 = \frac{1}{T_0} \int_{-T_0/2}^{T_0/2} x(t)\,dt \tag{4.7}$$

Thus we say that a_0 is the mean value of the periodic signal $x(t)$ over a single period.

FINDING THE OTHER COEFFICIENTS

We can obtain the remaining coefficients a_n by proceeding in the following manner: We multiply both sides of (4.3) by $\cos(2\pi mt/T_0)$ and then integrate over one period. Using (4.4) and (4.5), we see that the only term that survives is

$$a_n = \frac{1}{T_0} \int_{-T_0/2}^{T_0/2} x(t) \cos\left(\frac{2\pi nt}{T_0}\right) dt \tag{4.8}$$

Following a similar procedure but instead multiplying by $\sin(2\pi mt/T_0)$ and then calling on (4.6), we find that the remaining coefficients are given by

$$b_n = \frac{1}{T_0} \int_{-T_0/2}^{T_0/2} x(t) \sin\left(\frac{2\pi nt}{T_0}\right) dt \tag{4.9}$$

THE DIRICHLET CONDITIONS

The Dirichlet conditions tell us whether or not a given periodic function $x(t)$ can be represented by a Fourier series. These conditions are as follows:

- $x(t)$ is single valued over the fundamental period.
- $x(t)$ has a finite number of discontinuities, minima, and maxima over the fundamental period.
- $\int_{-T_0/2}^{T_0/2} |x(t)|\,dt < \infty$.

In regard to the last condition, we say that $x(t)$ is *absolutely integrable*.

We now proceed to find the Fourier series representations of a few functions. In the following examples, if a function does not appear to be periodic, assume

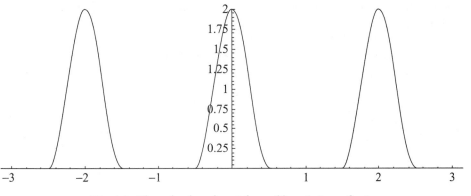

Fig. 4-1. The raised cosine pulse, $x(t) = 1 + \cos 2\pi t$.

that it has been duplicated up and down across the real line. (We show an explicit plot of this in Example 4-1.)

EXAMPLE 4-1
Find the Fourier series representation of the raised cosine pulse, as shown in Fig. 4-1. The signal is given by $x(t) = 1 + \cos 2\pi t$.

SOLUTION 4-1
First, we look at the plot to make some observations. The peak of the function is shown at $t = -2, t = 0, t = 2, \ldots$ and so we see that the fundamental period is $T_0 = 2$.

Looking at the figure, notice that this is an *even* function. Recall that $\cos x$ is even and $\sin x$ is odd. Looking at the formula for the Fourier series (4.3), we can see that if a function is even, the odd terms in the series should vanish. More specifically, this translates into saying that $b_n = 0$ for all n. Let's restate the formula here:

$$b_n = \frac{1}{T_0} \int_{-T_0/2}^{T_0/2} x(t) \sin\left(\frac{2\pi nt}{T_0}\right) dt$$

If $x(t)$ is even, then the function in the integrand is even $\cdot$ odd = odd, and the integral of an odd function over an interval that is symmetric about the origin vanishes—so these terms do not contribute to the Fourier series.

On the other hand, if the function is odd, then we have odd $\cdot$ odd = even in the case of the integrand for b_n and so these terms will not be zero. However, for the other terms, we have

$$a_n = \frac{1}{T_0} \int_{-T_0/2}^{T_0/2} x(t) \cos\left(\frac{2\pi nt}{T_0}\right) dt$$

which means that we are integrating odd · even = odd, and so $a_n = 0$ for all n. In the present example, we noted that the function at hand is even. So, we need to calculate only the a_n terms.

We begin by calculating a_0. Using (4.7), we have

$$a_0 = \frac{1}{2} \int_{-1}^{1} (1 + \cos 2\pi t) \, dt$$

However, looking at Fig. 4-1, notice that we need to integrate only over the range $-1/2 \leq t \leq 1/2$ since $x(t)$ is zero otherwise. So we find

$$a_0 = \frac{1}{2} \int_{-1/2}^{1/2} (1 + \cos 2\pi t) \, dt = \frac{1}{2} \int_{-1/2}^{1/2} dt + \frac{1}{2} \int_{-1/2}^{1/2} \cos 2\pi t \, dt$$

The first term can be evaluated immediately:

$$\frac{1}{2} \int_{-1/2}^{1/2} dt = \frac{1}{2} t \Big|_{-1/2}^{1/2} = \frac{1}{2} \left[\frac{1}{2} - \left(-\frac{1}{2}\right) \right] = \frac{1}{2} \left(\frac{1}{2} + \frac{1}{2} \right) = \frac{1}{2}$$

For the second term, we find

$$\frac{1}{2} \int_{-1/2}^{1/2} \cos 2\pi t \, dt = \frac{1}{4\pi} \sin 2\pi t \Big|_{-1/2}^{1/2} = \frac{1}{4\pi} (2 \sin \pi) = 0$$

So we conclude that $a_0 = 1/2$. Now let's calculate a_1. Using (4.8), we have

$$a_1 = \frac{1}{2} \int_{-1/2}^{1/2} (1 + \cos 2\pi t) \cos \pi t \, dt$$

$$= \frac{1}{2} \int_{-1/2}^{1/2} \cos \pi t \, dt + \frac{1}{2} \int_{-1/2}^{1/2} \cos 2\pi t \cos 2\pi t \, dt$$

Let's look at the first integral. We have

$$\frac{1}{2} \int_{-1/2}^{1/2} \cos \pi t \; dt = \frac{1}{2\pi} \sin \pi t \Big|_{-1/2}^{1/2} = \frac{1}{2\pi} \left(2 \sin \frac{\pi}{2} \right) = \frac{1}{\pi}$$

where we used $\sin(-x) = -\sin x$ when evaluating the result at the lower limit of integration. Looking at the second integral, we might be tempted to use (4.5) and so say it vanishes, since $m \neq n$ in this case (i.e., we are integrating $\cos 2\pi t \cos \pi t$). However, we *are not integrating over the period*, and so we cannot apply the orthogonality relations. Instead, we use the trigonometric identity $\cos 2\pi t = 1 - 2 \sin^2 \pi t$ to rewrite the integral as follows:

$$\frac{1}{2} \int_{-1/2}^{1/2} \cos 2\pi t \cos \pi t \; dt = \frac{1}{2} \int_{-1/2}^{1/2} (1 - 2 \sin^2 \pi t) \cos \pi t \; dt$$

$$= \frac{1}{2} \int_{-1/2}^{1/2} \cos \pi t \; dt - \int_{-1/2}^{1/2} (\sin^2 \pi t) \cos \pi t \; dt$$

The first term is identical to the integral we just calculated, and so it evaluates to $1/\pi$. We now consider the second term. If we take $u = \sin \pi t$ then $du = \pi \cos \pi t \; dt$ and we obtain

$$\int_{-1/2}^{1/2} (\sin^2 \pi t) \cos \pi t \; dt = \frac{1}{\pi} \int u^2 \; du = \frac{1}{\pi} \frac{u^3}{3}$$

$$= \frac{1}{3\pi} \sin^3 \pi t \Big|_{-1/2}^{1/2} = \frac{1}{3\pi} \left(2 \sin^3 \frac{\pi}{2} \right) = \frac{2}{3\pi}$$

Putting all of these results together, we obtain

$$a_1 = \frac{1}{\pi} + \frac{1}{\pi} - \frac{2}{3\pi} = \frac{4}{3\pi}$$

Let's calculate one more term. For a_2, (4.8) gives

$$a_2 = \frac{1}{2} \int_{-1/2}^{1/2} (1 + \cos 2\pi t) \cos 2\pi t \; dt$$

$$= \frac{1}{2} \int_{-1/2}^{1/2} \cos 2\pi t \; dt + \frac{1}{2} \int_{-1/2}^{1/2} \cos^2 2\pi t \; dt$$

Notice that with each term a_n, the integrand picks up a term $\cos(n\pi t)$. Proceeding, for the first term, we find

$$\frac{1}{2} \int_{-1/2}^{1/2} \cos 2\pi t \; dt = \frac{1}{4\pi} \sin 2\pi t \Big|_{-1/2}^{1/2} = \frac{1}{4\pi}[\sin \pi - \sin(-\pi)] = 0$$

To integrate the second term, we use the trigonometric identity

$$\cos^2 x = \frac{1 + \cos 2x}{2}$$

and the result is

$$\frac{1}{2} \int_{-1/2}^{1/2} \cos^2 2\pi t \; dt = \frac{1}{2} \int_{-1/2}^{1/2} \frac{1 + \cos 4\pi t}{2} \; dt$$

$$= \frac{1}{4} \int_{-1/2}^{1/2} dt + \frac{1}{4} \int_{-1/2}^{1/2} \cos 4\pi t \; dt$$

The second integral in this expression vanishes. The first term is elementary and turns out to be

$$\frac{1}{4} \int_{-1/2}^{1/2} dt = \frac{1}{4} t \Big|_{-1/2}^{1/2} = \frac{1}{4}\left(\frac{1}{2} + \frac{1}{2}\right) = \frac{1}{4}$$

and so we conclude that $a_2 = 1/4$. In the Quiz, we will show that all the b_n 's vanish. Therefore, using our results together with (4.3), we find that the Fourier series representation of the raised cosine pulse is

$$x(t) = \frac{1}{2} + \frac{8}{3\pi} \cos \pi t + \frac{1}{2} \cos 2\pi t + \cdots$$

EXAMPLE 4-2
Find the Fourier series of $x(t) = t^2$, $-1 < t < 1$.

SOLUTION 4-2
The function is shown in Fig. 4-2. Again, notice that this is an even function and so $b_n = 0$ for all n.

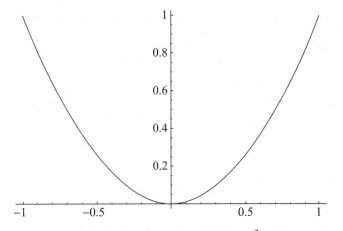

Fig. 4-2. We seek the Fourier series of $x(t) = t^2$, $-1 < t < 1$.

The period $T_0 = 2$ and so for the mean value of the function, we have

$$a_0 = \frac{1}{2} \int_{-1}^{1} t^2 \, dt = \frac{1}{6} t^3 \Big|_{-1}^{1} = \frac{1}{6}[1 - (-1)] = \frac{1+1}{6} = \frac{1}{3}$$

Next calculating a_1 using (4.8), we have

$$a_1 = \frac{1}{2} \int_{-1}^{1} t^2 \cos \pi t \, dt$$

We can do this integral using integration by parts. Recalling the formula $\int u \, dv = uv - \int v \, du$, if we let $u = t^2$ then we have $du = 2t \, dt$. Next we take $dv = \cos \pi t \, dt, \Rightarrow v = \frac{1}{\pi} \sin \pi t$ and so

$$a_1 = \frac{1}{2} \int_{-1}^{1} t^2 \cos \pi t \, dt = \frac{1}{2} \left[\frac{t^2 \sin \pi t}{\pi} \Big|_{-1}^{1} - \frac{2}{\pi} \int_{-1}^{1} t \sin \pi t \, dt \right]$$

Now,

$$\frac{t^2 \sin \pi t}{\pi} \Big|_{-1}^{1} = \frac{\sin \pi}{\pi} - \left(\frac{\sin(-\pi)}{\pi} \right) = 2\frac{\sin \pi}{\pi} = 0$$

so we are left with

$$a_1 = -\frac{1}{\pi} \int_{-1}^{1} t \sin \pi t \, dt$$

Once again, we use integration by parts. Setting $u = t \Rightarrow du = dt$ and

$$dv = \sin \pi t \, dt \Rightarrow v = -\frac{1}{\pi} \cos \pi t$$

we find

$$a_1 = -\frac{1}{\pi} \int_{-1}^{1} t \sin \pi t \, dt = -\frac{1}{\pi} \left[-\frac{t \cos \pi t}{\pi} \bigg|_{-1}^{1} + \frac{1}{\pi} \int_{-1}^{1} \cos \pi t \, dt \right]$$

The last integral vanishes, since

$$\frac{1}{\pi} \int_{-1}^{1} \cos \pi t \, dt = \frac{1}{\pi^2} \sin \pi t \bigg|_{-1}^{1} = \frac{1}{\pi^2} \sin \pi + \frac{1}{\pi^2} \sin \pi = 0$$

Therefore,

$$a_1 = -\frac{1}{\pi} \left[-\frac{t \cos \pi t}{\pi} \bigg|_{-1}^{1} \right] = \frac{\cos \pi}{\pi^2} - \left(-\frac{\cos \pi}{\pi^2} \right) = 2\frac{\cos \pi}{\pi^2} = -\frac{2}{\pi^2}$$

The other integrals in the series are similar and can be done using the same techniques. We can verify that

$$a_2 = \frac{1}{2} \int_{-1}^{1} t^2 \cos 2\pi t \, dt = \frac{1}{2\pi^2}$$

$$a_3 = \frac{1}{2} \int_{-1}^{1} t^2 \cos 3\pi t \, dt = -\frac{2}{9\pi^2}$$

Using these results, we find that

$$x(t) = a_0 + 2 \sum_{n=1}^{\infty} a_n \cos \pi n t$$

$$= \frac{1}{3} + 2\left(-\frac{2}{\pi^2}\cos\pi t + \frac{1}{2\pi^2}\cos 2\pi t - \frac{2}{9\pi^2}\cos 3\pi t + \cdots\right)$$

$$= \frac{1}{3} - \frac{4}{\pi^2}\cos\pi t + \frac{1}{\pi^2}\cos 2\pi t - \frac{4}{9\pi^2}\cos 3\pi t + \cdots$$

Complex Fourier Series

Using Euler's formulas, the cosine and sine functions can be written in terms of complex exponentials in the following way:

$$\cos x = \frac{e^{jx} + e^{-jx}}{2} \quad \text{and} \quad \sin x = \frac{e^{jx} - e^{-jx}}{2j}$$

These formulas allow us to write (4.3) in a more compact form involving complex exponentials. For notational convenience, for the time being, let's write $x = 2\pi t / T_0$. Then we have

$$x(t) = a_0 + 2\sum_{n=1}^{\infty}\left[a_n\cos(nx) + b_n\sin(nx)\right]$$

$$= a_0 + 2\sum_{n=1}^{\infty}\left[a_n\left(\frac{e^{jnx} + e^{-jnx}}{2}\right) + b_n\left(\frac{e^{jnx} - e^{-jnx}}{2j}\right)\right]$$

Now let's group terms together by getting everything that multiplies e^{+jnx} together in one term and everything that multiplies e^{-jnx} in the other term. When we do that, we arrive at

$$x(t) = a_0 + \sum_{n=1}^{\infty}\left[(a_n - jb_n)e^{jnx} + (a_n + jb_n)e^{-jnx}\right]$$

To make this step, note that we also used $1/j = -j$. Now we make the following definitions:

$$c_0 = a_0, \quad c_n = a_n - jb_n, \quad \text{and} \quad c_{-n} = a_n + jb_n \qquad (4.10)$$

With this change, the series becomes

$$x(t) = c_0 + \sum_{n=1}^{\infty}\left[c_n e^{jnx} + c_{-n}e^{-jnx}\right]$$

We can put everything together into one summation. Recalling that we had defined $x = 2\pi t/T_0$, we obtain the *complex exponential Fourier series* representation

$$x(t) = \sum_{-\infty}^{\infty} c_n \exp\left(j\frac{2\pi nt}{T_0}\right) \tag{4.11}$$

The coefficients, which as we have seen can be complex numbers, are given by

$$c_n = \frac{1}{T_0} \int_{-\infty}^{\infty} x(t) e^{-j2\pi nt/T_0} \, dt \tag{4.12}$$

Since the coefficients are complex numbers, we can write them in the polar representation as

$$c_n = |c_n| e^{j\phi_n} \tag{4.13}$$

where $\phi_n = \arg(c_n)$ is the phase. It is common to plot the amplitude and phase against frequency. A plot of $|c_n|$ against frequency is known as the *amplitude spectrum*, while a plot of ϕ_n against frequency is known as the *phase spectrum* of $x(t)$.

Power in Periodic Signals

The average power of a periodic signal $x(t)$ over one period is given by

$$P = \frac{1}{T_0} \int_{-T_0/2}^{T_0/2} |x(t)|^2 \, dt$$

It can be shown (Parseval's theorem) that if we represent $x(t)$ using (4.11) then we can write the power as

$$P = \frac{1}{T_0} \int_{-T_0/2}^{T_0/2} |x(t)|^2 dt = \sum_{n=-\infty}^{\infty} |c_n|^2 \tag{4.14}$$

EXAMPLE 4-3
Find the complex exponential Fourier representation of $x(t) = 2\sin t \cos t$.

SOLUTION 4-3
Very little calculation is necessary in this case. We have

$$x(t) = 2\sin t \cos t = \sin 2t = \frac{e^{j2t} - e^{-j2t}}{2j} = -j\frac{1}{2}e^{j2t} + je^{-j2t}$$

The period of $x(t)$ is π, and so

$$\frac{2\pi nt}{T_0} = \frac{2\pi nt}{\pi} = 2nt$$

Looking at (4.11), we conclude that

$$c_0 = 0, \quad c_1 = -j\frac{1}{2}, \quad c_{-1} = j\frac{1}{2}$$

and all other terms are zero.

EXAMPLE 4-4
Find the complex exponential Fourier series representation of

$$x(t) = \frac{4}{\pi}(\sin t + \sin 3t)$$

SOLUTION 4-4
A plot of the function over one period is shown in Fig. 4-3.
   We write each term separately and use Euler's formulas to write the sine functions in terms of exponentials:

$$x(t) = \frac{4}{\pi}(\sin t + \sin 3t)$$

$$= \frac{4}{\pi}\sin t + \frac{4}{\pi}\sin 3t$$

$$= \frac{4}{\pi}\left(\frac{e^{jt} - e^{-jt}}{2j}\right) + \frac{4}{\pi}\left(\frac{e^{j3t} - e^{-j3t}}{2j}\right)$$

$$= -\frac{2}{j3\pi}e^{-j3t} - \frac{2}{j\pi}e^{-jt} + \frac{2}{j\pi}e^{jt} + \frac{2}{j3\pi}e^{j3t}$$

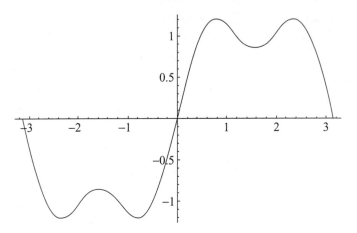

Fig. 4-3. The signal $x(t) = 4/\pi(\sin t + \sin 3t)$.

The formula for the complex exponential Fourier series is

$$x(t) = \sum_{-\infty}^{\infty} c_n \exp\left(j\frac{2\pi nt}{T_0}\right)$$

Using the period $T_0 = 2\pi$, we see that in this case we can write $x(t) = \sum_{-\infty}^{\infty} c_n \exp(jnt)$, and so we find that

$$c_1 = \frac{2}{j\pi} = -c_{-1}, \qquad c_3 = \frac{2}{j3\pi} = -c_{-3}$$

and all other terms in the series are zero.

EXAMPLE 4-5
Find the complex exponential Fourier series representation of the periodic impulse train, defined by

$$\delta_T(t) = \sum_{n=-\infty}^{\infty} \delta(t - nT)$$

SOLUTION 4-5
The periodic impulse train is a series of impulse functions separated by the period T. This is shown schematically in Fig. 4-4.

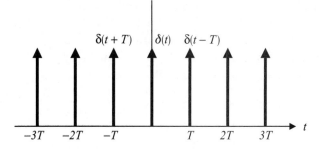

Fig. 4-4. A schematic representation of a periodic impulse train with period T.

The complex exponential Fourier series representation of a signal is given by (4.11), which we reproduce here for convenience.

$$x(t) = \sum_{-\infty}^{\infty} c_n \exp\left(j\frac{2\pi nt}{T}\right)$$

The coefficients in this case are given by

$$c_n = \frac{1}{T} \int_{-\infty}^{\infty} \delta(t) e^{-j2\pi nt/T}\, dt = \frac{1}{T}$$

where we have used the sampling property of the delta function. This means that the complex exponential series representation of the impulse train is

$$\delta_T(t) = \frac{1}{T} \sum_{n=-\infty}^{\infty} \exp\left(j\frac{2\pi nt}{T}\right)$$

The Fourier Transform

We now turn to the central topic of this chapter, the *Fourier transform*. This is a mathematical tool that will allow us to extend the basic idea of the Fourier series to incorporate nonperiodic signals. Think of it as a kind of generalization of the Fourier series concept—because it will allow us to obtain a frequency representation of periodic and nonperiodic signals.

In short, the Fourier transform lets us convert the representation of a signal in time into a representation of the signal in frequency. The *inverse Fourier transform* does the opposite—it lets us convert the representation of a signal in frequency into a representation of the signal in time.

It is typical to use a lower case letter to denote a signal in the time domain. We have been denoting such signals by $x(t)$, indicating that x is some continuous function of the time variable t. We denote the corresponding function of frequency by an upper case letter; for example, we can write $X(f)$. A function of angular frequency is written as $X(\omega)$. While we can convert easily between functions of frequency and angular frequency by using the relation $\omega = 2\pi f$, we will stick to using plain frequency because the unit of f is hertz, which as readers know is the unit commonly used when discussing communications signals. Just keep in mind that it is possible to describe the Fourier transform in terms of ω and is commonly done in other applications.

We aren't going to worry about the derivation of the Fourier transform, which is relatively simple and can be found in any signals textbook. Instead, we are just going to state what it is and then start finding out how to calculate and use them.

The *Fourier transform* of a $x(t)$ signal is given by

$$X(f) = \int_{-\infty}^{\infty} x(t) \exp\left(-j2\pi ft\right) dt \qquad (4.15)$$

Therefore the Fourier transform converts a time-domain representation of a signal into its frequency representation. Going in the reverse direction, we obtain the *inverse Fourier transform*

$$x(t) = \int_{-\infty}^{\infty} X(f) \exp\left(j2\pi ft\right) df \qquad (4.16)$$

In order for us to be able to find the Fourier transform of a given signal $x(t)$, it is sufficient that this signal satisfy the *Dirichlet conditions*:

- Within a finite time interval, $x(t)$ is single valued.
- $x(t)$ is absolutely integrable, meaning $\int_{-\infty}^{\infty} |x(t)| dt < \infty$.
- Within a finite time interval, $x(t)$ has a finite number of minima and maxima.
- Within a finite time interval, $x(t)$ has a finite number of discontinuities, and these discontinuities are finite.

The Dirichlet conditions are *sufficient* but not *strictly necessary*. What this means is that if the Dirichlet conditions are satisfied by $x(t)$, this guarantees that $x(t)$ has a Fourier transform. However if they are not satisfied for some signal

$y(t)$, then the Fourier transform might still exist anyway. The perfect example of this is the unit impulse function.

EXAMPLE 4-6
Find the Fourier transforms of $\delta(t)$ and $\delta(t - a)$.

SOLUTION 4-6
We simply apply the formula given in (4.15) and use the sampling property of the unit impulse function. In the first case, we find that

$$X_1(f) = \int_{-\infty}^{\infty} \delta(t) \exp(-j2\pi ft)\, dt = 1$$

We obtained this result by simply evaluating $\exp(-j2\pi ft)$ at $t = 0$. In the other case, we have

$$X_2(f) = \int_{-\infty}^{\infty} \delta(t - a) \exp(-j2\pi ft)\, dt = \exp(-j2\pi fa)$$

FOURIER TRANSFORM PAIRS

There is a particular shorthand notation that is used frequently in many textbooks to denote the relationship between a given signal in the time domain and its Fourier transform. This is the *Fourier transform pair*. For a signal $x(t)$ and its Fourier transform $X(f)$, we write

$$x(t) \rightleftharpoons X(f) \tag{4.17}$$

For the impulse functions we saw in the last example, we have

$$\delta(t) \rightleftharpoons 1$$

$$\delta(t - a) \rightleftharpoons \exp(-j2\pi fa)$$

THE *SINC* FUNCTION

A function we are sure to see in the context of signal processing is the *sinc* function that is defined as

$$\text{sinc } t = \frac{\sin \pi t}{\pi t} \tag{4.18}$$

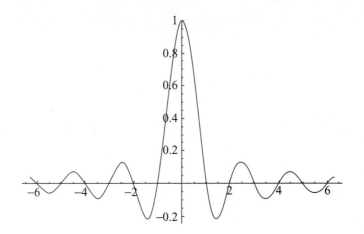

Fig. 4-5. The sinc function.

We can take the limit $t \to 0$ and apply L'Hopital's rule to see that this is in fact equal to unity at the origin. We show a plot of the function in Fig. 4-5.

EXAMPLE 4-7
Find the Fourier transform of a square pulse, as shown in Fig. 4-6.

SOLUTION 4-7
This pulse has the constant value A for $-b/2 < t < b/2$. Since it is nonzero only over a finite, symmetric range of t, the Fourier transform (4.15) is greatly simplified. We have

$$X(f) = A \int_{-b/2}^{b/2} \exp\left(-j2\pi ft\right) dt$$

This integral is pretty simple to do. We set $u = -j2\pi ft \Rightarrow du = -j2\pi f \, dt$. When $t = b/2$ we have $u = -j\pi fb$, while at the lower limit when $t = -b/2$

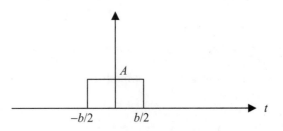

Fig. 4-6. A square pulse centered about the origin of height A.

we have $u = j\pi\, fb$. So the Fourier transform becomes

$$X(f) = A \int_{-b/2}^{b/2} \exp\left(-j2\pi ft\right) dt = -\frac{A}{j2\pi f} \int_{j\pi fb}^{-j\pi fb} e^u\, du$$

$$= \frac{A}{j2\pi f} \int_{-j\pi fb}^{j\pi fb} e^u\, du = \frac{A}{j2\pi f} e^u \Big|_{-j\pi fb}^{j\pi fb}$$

Evaluating at the upper and lower limits, we obtain

$$X(f) = \frac{A}{j2\pi f}\left(e^{j\pi fb} - e^{-j\pi fb}\right) = \frac{A}{\pi f}\left(\frac{e^{j\pi fb} - e^{-j\pi fb}}{2j}\right)$$

$$= A\frac{\sin \pi fb}{\pi f} = Ab\frac{\sin \pi fb}{\pi fb} = Ab\,\mathrm{sinc}\,(fb)$$

EXAMPLE 4-8
Find the Fourier transform of the triangular pulse, as shown in Fig. 4-7.

SOLUTION 4-8
With a little algebra we can find the functional form of this signal. We will con-sider two functions, $x_<(t)$ for $t < 0$ and $x_>(t)$ for $t > 0$. Since these functions are linear, they will be of the form $x(t) = mt + b$. When $t < 0$, we notice that $x_<(t) = 0$ when $t = -1$, giving us the relation

$$0 = -m + b \Rightarrow b = m$$

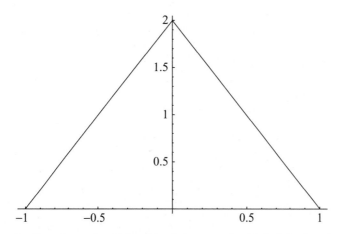

Fig. 4-7. A triangular pulse defined for $-1 < t < 1$ with height 2 at the origin.

At $t = 0$, we have $x_<(t) = 2$ and so

$$2 = b$$

Therefore, $x_<(t) = 2t + 2$. A similar exercise for $t > 0$ shows that $x_>(t) = -2t + 2$. Now we can use this information to split the defining integral of the Fourier transform up into two parts. Applying (4.15) in this situation, we obtain

$$X(f) = \int_{-\infty}^{\infty} x(t) \exp(-j2\pi ft) \, dt$$

$$= \int_{-1}^{0} (2t + 2) \exp(-j2\pi ft) \, dt + \int_{0}^{1} (-2t + 2) \exp(-j2\pi ft) \, dt$$

$$= X_<(f) + X_>(f)$$

The technique required to do both integrals will be the same, so let's calculate only the first integral explicitly. We want to find

$$\int_{-1}^{0} (2t + 2) \exp(-j2\pi ft) \, dt$$

We can do this using integration by parts. If we let $u = 2t + 2$, then $du = 2dt$. Next we choose $dv = e^{-j2\pi ft}$, giving

$$v = -\frac{e^{-j2\pi ft}}{j2\pi f}$$

The integration by parts formula tells us that

$$\int u \, dv = uv - \int v \, du$$

With our choices of u and v, we have

$$\int_{-1}^{0} (2t + 2) \exp(-j2\pi ft) \, dt = -\frac{(2t + 2)e^{-j2\pi ft}}{j2\pi f} \bigg|_{-1}^{0} + \frac{1}{j\pi f} \int_{-1}^{0} e^{-j2\pi ft} \, dt$$

Looking at the leading term, at the upper limit $t = 0$ and so

$$-\frac{(2t + 2)e^{-j2\pi ft}}{j2\pi f} = -\frac{1}{j\pi f}$$

At the lower limit, $t = -1$ and this term evaluates to

$$-\frac{(-2+2)e^{j2\pi f}}{j2\pi f} = 0$$

So we're left with

$$\int_{-1}^{0} (2t + 2) \exp(-j2\pi ft)\, dt = -\frac{1}{j\pi f} + \frac{1}{j\pi j} \int_{-1}^{0} e^{-j2\pi ft}\, dt$$

The last piece becomes

$$\int_{-1}^{0} e^{-j2\pi ft}\, dt = -\frac{1}{j2\pi f} e^{-j2\pi ft} \Big|_{-1}^{0} = -\frac{1}{j2\pi f} + \frac{e^{j2\pi f}}{j2\pi f}$$

Putting these results together, we have (remember to multiply the result on the above line by $1/j\pi f$)

$$X_<(f) = -\frac{1}{j\pi f} + \frac{1}{2\pi^2 f^2} - \frac{e^{j2\pi f}}{2\pi^2 f^2}$$

A similar exercise applied to $\int_0^1 (-2t + 2) \exp(-j2\pi ft)\, dt$ yields

$$X_>(f) = \frac{1}{j\pi f} + \frac{1}{2\pi^2 f^2} - \frac{e^{-j2\pi f}}{2\pi^2 f^2}$$

Adding these terms together, we obtain

$$X_<(f) + X_>(f) = -\frac{1}{j\pi f} + \frac{1}{2\pi^2 f^2} - \frac{e^{j2\pi f}}{2\pi^2 f^2}$$
$$- + \frac{1}{j\pi f} + \frac{1}{2\pi^2 f^2} - \frac{e^{-j2\pi f}}{2\pi^2 f^2}$$
$$= \frac{1}{\pi^2 f^2} - \frac{e^{j2\pi f}}{2\pi^2 f^2} - \frac{e^{-j2\pi f}}{2\pi^2 f^2}$$

Now let's use

$$\cos t = \frac{e^{jt} + e^{-jt}}{2}$$

to write this as

$$X(f) = \frac{1}{\pi^2 f^2} - \frac{e^{j2\pi f}}{2\pi^2 f^2} - \frac{e^{-j2\pi f}}{2\pi^2 f^2} = \frac{1}{\pi^2 f^2} - \frac{1}{\pi^2 f^2}\left(\frac{e^{j2\pi f} + e^{-j2\pi f}}{2}\right)$$

$$= \frac{1 - \cos 2\pi f}{\pi^2 f^2}$$

Now we use the trigonometric identity $\cos 2t = 1 - 2\sin^2 t$ together with the definition of the sinc function to obtain the Fourier transform of a triangular pulse

$$X(f) = \frac{1 - \cos 2\pi f}{\pi^2 f^2} = \frac{1 - \left(1 - 2\sin^2 \pi f\right)}{\pi^2 f^2}$$

$$= 2\frac{\sin^2 \pi f}{\pi^2 f^2} = 2\left(\frac{\sin \pi f}{\pi f}\right)^2 = 2\operatorname{sinc}^2 f$$

In general, the Fourier transform of a triangular pulse of height A defined over $-T < t < T$ is $AT^2 \operatorname{sinc}^2 (Tf)$.

EXAMPLE 4-9
Find the Fourier transform of $e^{-t} \cos 3t u(t)$.

SOLUTION 4-9
Using (4.15) along with the properties of the unit step function, the Fourier transform is given by

$$X(f) = \int_0^\infty e^{-t} \cos 3t \, e^{-j2\pi ft} \, dt$$

Now we use

$$\cos 3t = \frac{e^{j3t} + e^{-j3t}}{2}$$

to write this as

$$X(f) = \int_0^\infty e^{-t} \left(\frac{e^{j3t} + e^{-j3t}}{2}\right) e^{-j2\pi ft} \, dt$$

$$= \frac{1}{2}\int_0^\infty e^{-t} e^{j3t} e^{-j2\pi ft} \, dt + \frac{1}{2}\int_0^\infty e^{-t} e^{-j3t} e^{-j2\pi ft} \, dt$$

A little algebraic manipulation gives

$$X(f) = \frac{1}{2} \int_0^\infty e^{(-1+3j-j2\pi f)t} \, dt + \frac{1}{2} \int_0^\infty e^{-(1+3j+j2\pi f)t} \, dt$$

$$= \frac{1}{2(-1+3j-j2\pi f)} e^{(-1+3j-j2\pi f)t} \Big|_0^\infty$$

$$- \frac{1}{2(1+3j+j2\pi f)} e^{(1+3j+j2\pi f)t} \Big|_0^\infty$$

$$= \frac{-1}{2(-1+3j-j2\pi f)} + \frac{1}{2(1+3j+j2\pi f)}$$

$$= \frac{-1}{(-2+6j-j4\pi f)} + \frac{1}{(2+6j+j4\pi f)}$$

Cross-multiplication gives

$$X(f) = -\frac{2+6j+j4\pi f}{[-2+(6-4\pi f)j][2+(6+4\pi f)j]}$$

$$+ \frac{-2+6j-j4\pi f}{[-2+(6-4\pi f)j][2+(6+4\pi f)j]}$$

$$= \frac{-4-j8\pi f}{[-2+(6-4\pi f)j][2+(6+4\pi f)j]}$$

Looking at the denominator, we obtain

$$[-2+(6-4\pi f)j][2+(6+4\pi f)j]$$
$$= -4 - 2(6+4\pi f)j + 2(6-4\pi f)j - (6-4\pi f)(6+4\pi f)$$
$$= -4 - 12j - j8\pi f + 12j - j8\pi f - 36 + 16\pi^2 f^2$$
$$= -40 - j16\pi f + 16\pi^2 f^2$$
$$= -4\left(10 + j4\pi f - 4\pi^2 f^2\right)$$

Inserting this into the above result, we have

$$X(f) = \frac{-4-j8\pi f}{-4(10+j4\pi f - 4\pi^2 f^2)} = \frac{1+j2\pi f}{10+j4\pi f - 4\pi^2 f^2}$$

So we have the Fourier transform pair

$$e^{-t} \cos 3tu\,(t) \rightleftharpoons \frac{1 + j2\pi f}{10 + j4\pi f - 4\pi^2 f^2}$$

EXAMPLE 4-10
Show that the Fourier transform of a Gaussian pulse is a Gaussian in the frequency domain.

SOLUTION 4-10
A Gaussian pulse $x(t) = e^{-\pi t^2}$ is shown in Fig. 4-8.

To find the Fourier transform, we will need to use the following integral:

$$\int_{-\infty}^{\infty} e^{-at^2}\, e^{bt}\, dt = \sqrt{\frac{\pi}{a}} e^{b^2/4a} \tag{4.19}$$

Using (4.15), we have

$$X(f) = \int_{-\infty}^{\infty} e^{-\pi t^2}\, e^{-j2\pi ft}\, dt$$

Now applying (4.19), we identify that $a = \pi$ and $b = -j2\pi f$ and so

$$\frac{b^2}{4a} = \frac{(-j2\pi f)^2}{4\pi} = \frac{-4\pi^2 f^2}{4\pi} = -\pi f^2$$

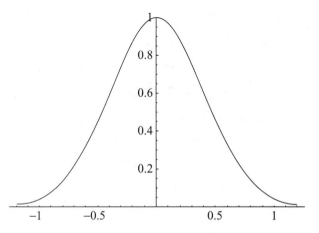

Fig. 4-8. A Gaussian pulse.

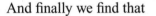

And finally we find that

$$X(f) = \int_{-\infty}^{\infty} e^{-\pi t^2} e^{-j2\pi ft} \, dt = e^{-\pi f^2}$$

PROPERTIES OF THE FOURIER TRANSFORM

In this next section, we will examine several of the basic properties of Fourier transforms. Many of these properties are useful in doing calculations and can save a great deal of work. We review each property in turn. In the following, we denote the Fourier transform operation by FT.

Time shifting

If a signal is shifted in the time domain by a, then its Fourier transform is multiplied by $e^{-j2\pi af}$. In other words,

$$\text{FT}\,[x(t-a)] = e^{-j2\pi af} X(f) \qquad (4.20)$$

Frequency shifting

Multiplication of a signal in the time domain by $e^{-j2\pi f_0 t}$ is equivalent to a shift in the frequency domain; that is, given the Fourier transform pair $x(t) \rightleftharpoons X(f)$,

$$e^{j2\pi f_0 t} x(t) \rightleftharpoons X(f - f_0) \qquad (4.21)$$

Time scaling

Compression in time of a signal $x(t)$ causes a broadening of frequency of $X(f)$, while broadening in time of $x(t)$ results in a compression in frequency of $X(f)$. Focusing on the first case, we can compress a signal in time by multiplying the time variable t by a positive constant a. If $x(t) \rightleftharpoons X(f)$ constitute a Fourier transform pair, then

$$x(at) \rightleftharpoons \frac{1}{|a|} X\left(\frac{f}{a}\right) \qquad (4.22)$$

To illustrate compression in the time domain, we compare $e^{-t}u(t)$ with $e^{-2t}u(2t)$ in Fig. 4-9. Notice that $e^{-2t}u(2t)$ is a smaller pulse, or compressed in time.

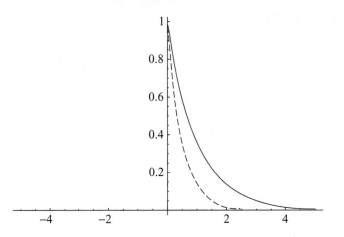

Fig. 4-9. Compression in the time domain. Solid line is $e^{-t}u(t)$, while the dashed line is $e^{-2t}u(2t)$.

In Fig. 4-10, we consider the Fourier transforms of each function. Specifically, we plot the amplitude $|X(f)|$ versus frequency in each case. The spreading out in frequency is evident; also, notice the smaller amplitude that is due to the factor of $1/2$ in

$$x(at) \rightleftharpoons \frac{1}{|a|} X\left(\frac{f}{a}\right)$$

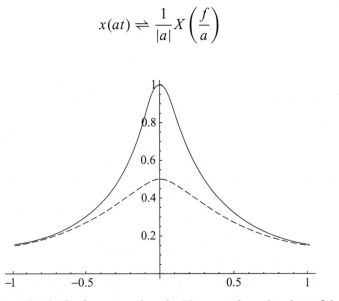

Fig. 4-10. Expansion in the frequency domain. Here, we show the plots of the Fourier transforms of $e^{-t}u(t)$ (the solid line) and $e^{-2t}u(2t)$ (the dashed line).

The superposition principle

Linearity of the Fourier transform follows immediately from the definition. Specifically,

$$ax_1(t) + bx_2(t) \rightleftharpoons aX_1(f) + bX_2(f) \tag{4.23}$$

Duality

The duality property is very useful when computing Fourier transforms, as we will see by example. The duality property tells us that if $x(t) \rightleftharpoons X(f)$, then it follows that

$$X(t) \rightleftharpoons x(-f) \tag{4.24}$$

Differentiation and integration

Looking at the inverse Fourier transform (4.16), consider what happens when we differentiate with respect to time. We obtain

$$\frac{dx}{dt} = \frac{d}{dt} \int_{-\infty}^{\infty} X(f) \exp(j2\pi ft) \, df$$

Since the derivative is with respect to time but the integration variable is frequency, we can pull the derivative operator inside the integral to give

$$\frac{dx}{dt} = \int_{-\infty}^{\infty} \frac{d}{dt} X(f) \exp(j2\pi ft) \, df = \int_{-\infty}^{\infty} j2\pi f X(f) \exp(j2\pi ft) \, df$$

This tells us that if $x(t) \rightleftharpoons X(f)$, then

$$\frac{dx}{dt} \rightleftharpoons j2\pi f X(f) \tag{4.25}$$

Now, if $X(0) = 0$, then we also have

$$\int_{-\infty}^{t} x(\tau) \, d\tau \rightleftharpoons \frac{1}{j2\pi f} X(f) \tag{4.26}$$

In summary, differentiation in the time domain corresponds to multiplication by $j2\pi f$ in the frequency domain, while integration in the time domain corresponds to division by $j2\pi f$ in the frequency domain.

Convolution

Convolution is certainly a big headache. Luckily the Fourier transform simplifies this for us. First, we make the unpleasant observation that multiplication of two functions in the time domain results in convolution in the frequency domain:

$$x_1(t)x_2(t) \rightleftharpoons \int_{-\infty}^{\infty} X_1(\sigma)X_2(f-\sigma)\,d\sigma \qquad (4.27)$$

On a more upbeat note, convolution in the time domain translates into multiplication in the frequency domain, i.e.,

$$\int_{-\infty}^{\infty} x_1(t)x_2(t-\tau)\,d\tau \rightleftharpoons X_1(f)X_2(f) \qquad (4.28)$$

This result is deemed so important (Well convolution is a pain, isn't it?) that it has been dubbed as the *convolution theorem*.

EVEN AND ODD PARTS OF A FUNCTION

Suppose that $x(t)$ is a real signal and we have decomposed it into its even and odd parts, i.e., $x(t) = x_e(t) + x_o(t)$. Furthermore, suppose that $X(f) = A(f) + jB(f)$. Then the following results hold

$$x_e(t) \rightleftharpoons A(f) \qquad (4.29)$$

$$x_o(t) \rightleftharpoons B(f) \qquad (4.30)$$

$$X(-f) = X^*(f) \qquad (4.31)$$

In words, the *real* part of the Fourier transform and the even part of the signal constitute a Fourier transform pair, while the *imaginary* part of the Fourier transform and the odd part of the signal constitute a Fourier transform pair. Equation (4.31) is necessary and sufficient to determine that a signal $x(t)$ is real.

EXAMPLE 4-11
Find the Fourier transform of the sinc function.

SOLUTION 4-11
We start by proceeding ahead blindly. Once again calling on (4.15), we have

$$X(f) = \int_{-\infty}^{\infty} \text{sinc}\,(t)\,\exp\,(-j2\pi ft)\,dt = \int_{-\infty}^{\infty} \frac{\sin \pi t}{\pi t}\,\exp\,(-j2\pi ft)\,dt$$

This integral looks hopeless. Let's simplify things a bit by writing

$$\sin \pi t = \frac{e^{j\pi t} - e^{-j\pi t}}{2j}$$

Now we have

$$X(f) = \int_{-\infty}^{\infty} \left(\frac{e^{j\pi t} - e^{-j\pi t}}{2j} \right) \frac{\exp\,(-j2\pi ft)}{\pi t}\,dt$$

$$= \frac{1}{2j} \int_{-\infty}^{\infty} \frac{e^{j\pi t}}{\pi t}\,\exp\,(-j2\pi ft)\,dt - \frac{1}{2j} \int_{-\infty}^{\infty} \frac{e^{-j\pi t}}{\pi t}\,\exp\,(-j2\pi ft)\,dt$$

The two integrals here are the same except for some sign differences, and so let's just focus on the first one. First let's use $e^A e^B = e^{A+B}$ to get everything in one exponential.

$$\frac{1}{2j} \int_{-\infty}^{\infty} \frac{e^{j\pi t}}{\pi t}\,\exp\,(-j2\pi ft)\,dt = \frac{1}{2j} \int_{-\infty}^{\infty} \frac{1}{\pi t}\,\exp\,[j\pi\,(1 - 2f)\,t]\,dt$$

This still looks pretty ugly; many readers probably don't know offhand how to do this. But instead of worrying about that, let's recall from Example 4-7 that the Fourier transform of a rectangular pulse of height A and width b centered about the origin was $Ab\,\text{sinc}\,(fb)$. The duality property tells us that we can find the Fourier transform of the sinc function by calling on this result, instead of wasting time calculating it explicitly. It is common to denote a rectangular pulse of height A and width b centered about the origin by the compact notation $A\,\text{rect}\,(t/b)$. Example 4-7 told us that

$$A\,\text{rect}\left(\frac{t}{b}\right) \rightleftharpoons Ab\,\text{sinc}\,(bf)$$

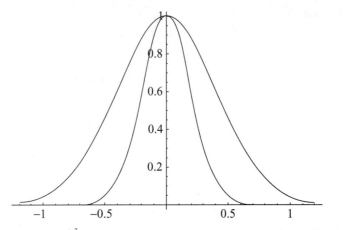

Fig. 4-11. The signal $e^{-\pi 4t^2}$, compressed in time, is the narrow pulse in this figure, while $x(t) = e^{-\pi t^2}$ extends out much further.

Using the duality property, we conclude that the Fourier transform of sinc (t) is rect (f); that is,

$$\text{sinc}(t) \rightleftharpoons \text{rect}(f)$$

EXAMPLE 4-12
Find the Fourier transform of $x(t) = e^{-\pi 4t^2}$.

SOLUTION 4-12
The time scaling factor here is 2, since $x(t) = e^{-\pi 4t^2} = e^{-\pi(2t)^2}$. In Fig. 4-11, this pulse is compared with $x(t) = e^{-\pi t^2}$.

Using the time scaling property, we know that since the Fourier transform of $x(t) = e^{-\pi t^2}$ is just $X(f) = e^{-\pi f^2}$, the Fourier transform of $x(2t) = e^{-\pi 4t^2}$ is given by

$$\frac{1}{2}X\left(\frac{f}{2}\right) = \frac{1}{2}e^{-\pi\frac{f^2}{4}}$$

Figure 4-12 compares the plots of the two Fourier transforms.

EXAMPLE 4-13
Find the Fourier transform of $x(t) = \cos 2\pi f_0 t$.

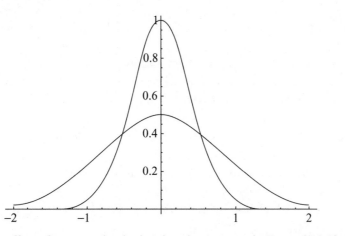

Fig. 4-12. The effect of compression in time is quite apparent in Example 4-12. Here, we see plots versus frequency of the Fourier transform of $x(t) = e^{-\pi t^2}$ (the narrow pulse) and the Fourier transform of $e^{-\pi 4t^2}$ (the flat, wide pulse). Compression in time has resulted in expansion in frequency.

SOLUTION 4-13
First, we write

$$x(t) = \cos 2\pi f_0 t = \frac{e^{j2\pi f_0 t} + e^{-j2\pi f_0 t}}{2}$$

then

$$X(f) = \int_{-\infty}^{\infty} \left(\frac{e^{j2\pi f_0 t} + e^{-j2\pi f_0 t}}{2} \right) e^{-j2\pi f t} \, dt$$

$$= \frac{1}{2} \int_{-\infty}^{\infty} e^{j2\pi (f_0 - f)t} \, dt + \frac{1}{2} \int_{-\infty}^{\infty} e^{-j2\pi (f_0 + f)t} \, dt$$

In Example 4-6, we found that the Fourier transform of the unit impulse was given by $X_2(f) = \int_{-\infty}^{\infty} \delta(t - a) \exp(-j2\pi f t) \, dt = \exp(-j2\pi f a)$. Now we use the duality property to deduce that the Fourier transform of $e^{j2\pi f_0 t}$ is a unit impulse in frequency. We find that

$$X(f) = \frac{1}{2} [\delta(f - f_0) + \delta(f + f_0)]$$

Spectrum Plots

We close out the chapter with an examination of spectrum plots. In general, the Fourier transform of a signal $x(t)$ is a *complex* function. This means that we can write $X(f)$ in polar representation; that is,

$$X(f) = |X(f)| e^{j\phi} \qquad (4.32)$$

where $|X(f)|$ is the amplitude of $X(f)$ and $\phi = \arg(X(f))$ is the phase of $X(f)$. A plot of $|X(f)|$ is known as the *amplitude spectrum* of the signal and a plot of $\phi = \arg(X(f))$ is the phase spectrum of the signal.

EXAMPLE 4-14
Suppose that the Fourier transform of some signal is

$$X(f) = \frac{1}{2 + jf}$$

Plot the amplitude and phase spectra.

SOLUTION 4-14
The magnitude of a complex number z is given by

$$|z|^2 = zz^*$$

The complex conjugate of the function is found by letting $j \to -j$, giving

$$X^*(f) = \frac{1}{2 - jf}$$

Therefore,

$$X(f)X^*(f) = \frac{1}{2 + jf}\frac{1}{2 - jf} = \frac{1}{4 + j2f - j2f + f^2} = \frac{1}{4 + f^2}$$

We need to take the square root to obtain $|X(f)|$:

$$|X(f)| = \frac{1}{\sqrt{4 + f^2}}$$

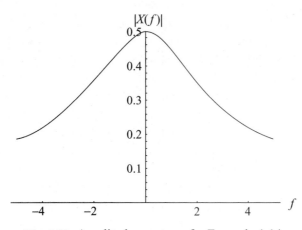

Fig. 4-13. Amplitude spectrum for Example 4-14.

A plot of this function versus frequency gives the amplitude spectrum. As shown in Fig. 4-13, this gives a bell-shaped curve.

To find the phase spectrum, recall that given a complex number $z = x + iy$, $\phi = \arg(z) = \arctan(y/x)$. First, we manipulate the function to get it in the form $z = x + iy$ as

$$\frac{1}{2 + jf} = \frac{1}{2 + jf}\left(\frac{2 - jf}{2 - jf}\right) = \frac{2 - jf}{4 + f^2}$$

This allows us to make the following identifications:

$$x = \frac{2}{4 + f^2} \quad \text{and} \quad y = \frac{-f}{4 + f^2}$$

So we find that

$$\frac{y}{x} = -\frac{f}{2}$$

The argument is then

$$\phi = \arg[X(f)] = \arctan(-f/2)$$

The phase spectrum is a plot of this function against frequency. In the present case, we have the result shown in Fig. 4-14.

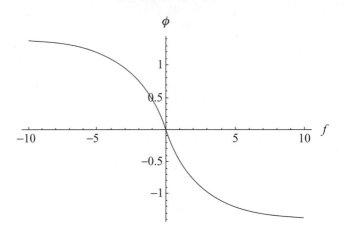

Fig. 4-14. An example phase spectrum, for $X(f) = 1/(2 + jf)$.

Parseval's Theorem

Parseval's theorem tells us that the energy content of a signal is equivalent to the *energy spectral density* of the signal, which is found by integrating $|X(f)|^2$. We quantify this statement by writing

$$\int_{-\infty}^{\infty} |x(t)|^2 \, dt = \frac{1}{2\pi} \int_{-\infty}^{\infty} |X(\omega)|^2 \, d\omega \qquad (4.33)$$

In the next chapter, we will use the mathematical tools developed in this chapter to work with signals in the frequency domain.

Quiz

1. Complete Example 4-1 by showing that $b_n = 0$.
2. Consider Example 4-2 and show that

$$a_3 = \frac{1}{2} \int_{-1}^{1} t^2 \cos 3\pi t \, dt = -\frac{2}{9\pi^2}$$

3. Find the Fourier series of $x(t) = t$, $-1 < t < 1$.
4. Find the complex exponential Fourier series of $x(t) = \cos 2t + \sin 8t$.
5. Find the Fourier transform of $x(t) = e^{-|t|} \sin 2t$.
6. Find the Fourier transform of $x(t) = e^{-|t|}$.

7. Find the Fourier transform of $x(t) = e^{-t}u(t)$.
8. Find the Fourier Transform of

$$x(t) = \begin{cases} 1 & t > 0 \\ -1 & t < 0 \end{cases}$$

9. Find the Fourier transform of $x(t) = u(t - 1)$.
10. Find the Fourier transform of $x(t) = e^{-(t-2)}u(t - 2)$.

CHAPTER 5

Some Applications of Fourier Analysis and Filters

In the last chapter we saw how to take any signal (periodic or not) and obtain its frequency content using the tools of Fourier analysis. Now let's explore the ways that the Fourier transform can be applied to signal processing. We begin by studying the frequency response of a system.

Frequency Response

In the last chapter we noted that the Fourier transform had one very nice property—it allowed us to convert convolution in the time domain into multiplication in the frequency domain. For most people convolution is a major pain, and so if we can find or look up the Fourier transform of the two signals then this property is a major plus. In short, the Fourier transform, which at first might seem to be a lot of extra work, turns out to be a very nice friend by giving us an easy way to find results involving convolution.

The reader will recall from Chapter 2 that given the impulse response $h(t)$ of a continuous linear time-invariant (LTI) system, the response $y(t)$ to any input signal $x(t)$ can be found by computing the convolution of the input signal and the impulse response of the system. That is

$$y(t) = x(t)^*h(t)$$

We aren't going to bother writing down the convolution integral in this chapter because the Fourier transform gives us a means to ignore it. The convolution theorem (4.28) allows us to write this expression in the frequency domain as

$$Y(f) = X(f)H(f) \tag{5.1}$$

We say that $H(f)$ is the *frequency response* or transfer function of the system. This is indicated schematically in Fig. 5-1.

We recall that in general the Fourier transform is a complex function of frequency. This means that we can write the frequency response in polar form

$$H(f) = |H(f)|e^{j\theta_H} \tag{5.2}$$

where the magnitude of the frequency response $|H(f)|$ is known as the *amplitude response* of the signal and the argument $\theta_H(f) = \arg[H(f)]$ is called the *phase response*.

Now, given (5.1), we can write the Fourier transform of the output signal in the following way:

$$y(t) = \int_{-\infty}^{\infty} Y(f)e^{j2\pi ft}\, df = \int_{-\infty}^{\infty} X(f)H(f)e^{j2\pi ft}\, df$$

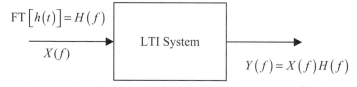

Fig. 5-1. A schematic representation of the role of frequency response in determining the output given an arbitrary input signal $X(f)$.

Basically this tells us that the response in the frequency domain of a continuous time LTI system is completely determined by $H(f)$. Now let's get some more information about the relationship between the response $Y(f)$ and the input $X(f)$ using (5.1) and writing each function in polar form. We have

$$Y(f) = |Y(f)|e^{j\theta_y}, \qquad X(f) = |X(f)|e^{j\theta_x}, \qquad H(f) = |H(f)|e^{j\theta_H}$$

Recalling that $e^f e^g = e^{f+g}$ and writing the right side of (5.1) in polar form gives us

$$X(f)H(f) = (|X(f)| e^{j\theta_x})(|H(f)| e^{j\theta_H})$$
$$= |X(f)||H(f)| e^{j\theta_x} e^{j\theta_H} = |X(f)||H(f)| e^{j(\theta_x+\theta_H)}$$

This tells us that the amplitude of the output signal is related to the product of the amplitudes of the input signal and the frequency response

$$|Y(f)| = |X(f)||H(f)| \tag{5.3}$$

On the other hand, the phase of the output signal is the sum of the phases of the input signal and frequency response

$$\theta_Y(f) = \theta_X(f) + \theta_H(f) \tag{5.4}$$

This isn't surprising; it's just a basic fact involving the arithmetic of complex numbers. But remember, when finding the output in the frequency domain, amplitudes multiply and phases add.

EXAMPLE 5-1
The operation of a continuous time LTI system is described by the differential equation

$$\frac{dy}{dt} + 3y(t) = 2x(t)$$

Find the output $y(t)$ if the input signal is given by $x(t) = e^{-t}u(t)$. The input signal is shown in Fig. 5-2.

SOLUTION 5-1
Solving this equation in the time domain would not be very much fun. But recalling that the Fourier transform turns a derivative in the time domain into multiplication by $j2\pi f$ in the frequency domain (4.25), if we take the Fourier

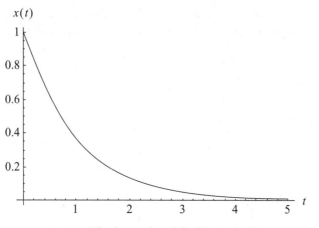

Fig. 5-2. The input signal for Example 5-1.

transform of the given differential equation we have

$$j2\pi f Y(f) + 3Y(f) = 2\frac{1}{1 + j2\pi f}$$

To obtain the right-hand side, we use the fact that the Fourier transform of the input signal $x(t) = e^{-t}u(t)$ is given by $X(f) = 1/(1 + j2\pi f)$. Now let's do a bit of algebraic manipulation to write the equation in the frequency domain as

$$Y(f)(3 + j2\pi f) = 2\frac{1}{1 + j2\pi f}$$

Dividing both sides by $(3 + j2\pi f)$ gives

$$Y(f) = \frac{2}{(3 + j2\pi f)}\frac{1}{1 + j2\pi f}$$

Now, before going any further recall (5.1). The expression we have here is nothing more than

$$Y(f) = H(f)X(f)$$

So the frequency response of this system is

$$H(f) = \frac{2}{(3 + j2\pi f)} = 2\frac{1}{3 + j2\pi f}$$

We note that the Fourier transform of $e^{-at}u(t)$ is given by $\frac{1}{a+j2\pi f}$. Looking at $H(f)$, we conclude that $a = 3$ and so the impulse response is

$$h(t) = 2\,e^{-3t}u(t)$$

Let's return to the task of finding the output signal. We want to manipulate $Y(f)$ keeping the Fourier transform pair

$$e^{-at}u(t) \rightleftharpoons \frac{1}{a + j2\pi f}$$

in the back of our minds. What we will do is try to rewrite $Y(f)$ so that we can get rid of the multiplication of two functions of frequency that we see here and turn that into a summation. We can do this by using the approach of partial fraction decomposition. That is we write

$$\frac{2}{(3 + j2\pi f)(1 + j2\pi f)} = \frac{A}{(3 + j2\pi f)} + \frac{B}{(1 + j2\pi f)}$$

where A and B are unknown constants to be determined. The first step is to multiply both sides by the product $(3 + j2\pi f)(1 + j2\pi f)$, which gives

$$2 = A(1 + j2\pi f) + B(3 + j2\pi f)$$

To solve for the value of each unknown constant, we fix f appropriately to eliminate one or the other term. First let's eliminate B. To do this we set

$$3 + j2\pi f = 0$$

Rearranging terms we have

$$j2\pi f = -3 \;\Rightarrow\; f = -\frac{3}{j2\pi} = j\frac{3}{2\pi}$$

Now we substitute this result into the other term. This gives

$$2 = A(1 + j2\pi f) = A\left[1 + j2\pi\left(j\frac{3}{2\pi}\right)\right] = A(1 - 3) = -2A$$

This allows us to conclude that $A = -1$. Next we follow the same procedure to find B. This time we set

$$1 + j2\pi f = 0 \implies f = -\frac{1}{j2\pi} = j\frac{1}{2\pi}$$

Substitution yields

$$2 = B(3 + j2\pi f) = B\left[3 + j2\pi\left(\frac{j}{2\pi}\right)\right] = B(3 - 1) = 2B$$

And so we have determined that $B = 1$. This allows us to write the Fourier transform of the output response as

$$Y(f) = -\frac{1}{(3 + j2\pi f)} + \frac{1}{(1 + j2\pi f)}$$

Now we have two expressions we can use to simply read off the inverse Fourier transform. Let's denote the inverse Fourier transform by FT^{-1}, then

$$\text{FT}^{-1}\left[-\frac{1}{(3 + j2\pi f)}\right] = -e^{-3t}u(t), \qquad \text{FT}^{-1}\left[\frac{1}{(1 + j2\pi f)}\right] = e^{-t}u(t),$$

So the output signal is

$$y(t) = -e^{-3t}u(t) + e^{-t}u(t) = u(t)\left(e^{-t} - e^{-3t}\right)$$

This output signal is shown in Fig. 5-3.

EXAMPLE 5-2
Given that

$$\frac{dy}{dt} = 2x(t)$$

Find $y(t)$ when $x(t) = \delta(t)$.

SOLUTION 5-2
From Appendix A we find the Fourier transform pair

$$\delta(t) \rightleftharpoons 1$$

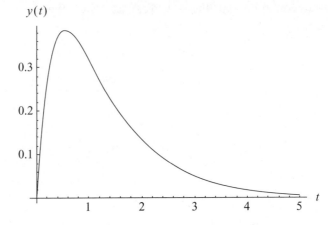

Fig. 5-3. The output signal for Example 5-1.

And so, taking the Fourier transform of the equation we obtain

$$j2\pi f Y(f) = 2$$

This leads to

$$Y(f) = \frac{1}{j\pi f}$$

Again consulting Appendix A, we see that

$$y(t) = \text{sgn}(t)$$

The Hilbert Transform

The *sgn* or signum function describes a *Hilbert transformer*. Specifically, the frequency response is defined by

$$H(f) = \begin{cases} e^{-j(\pi/2)} & f > 0 \\ e^{j(\pi/2)} & f < 0 \end{cases} \qquad (5.5)$$

Notice that $e^{j(\pi/2)} = \cos(\pi/2) + j\infty \sin(\pi/2) = j$. In the Quiz, you will show that the impulse response function in this case is $h(t) = 1/\pi t$. The output of the Hilbert transformer in the time domain is given by

$$y(t) = x(t)^* h(t) = \frac{1}{\pi} \int_{-\infty}^{\infty} \frac{x(\tau)}{t - \tau} \, d\tau \qquad (5.6)$$

Distortion

If transmission through an LTI system preserves signal shape, we say that the transmission is *distortionless*. While we require the shape of the signal to remain the same, we still have distortionless transmission even if the amplitude has changed or if there is a time delay. If we call an arbitrary time delay a then in the time domain distortionless transmission is described by

$$y(t) = Kx(t - a) \qquad (5.7)$$

Here K is called the *gain constant*. The time delay property of the Fourier transform tells us that

$$\text{FT}\,[x(t - a)] = e^{-j2\pi fa} X(f)$$

and so in the frequency domain, distortionless transmission is described by

$$Y(f) = K\,e^{-j2\pi fa} X(f) \qquad (5.8)$$

Now recall that we started the chapter with a description of the relationship between input and output in the frequency domain (5.1), which told us that

$$Y(f) = X(f)\,H(f)$$

Comparison with (5.8) leads us to conclude that in the case of distortionless transmission, the frequency response is

$$H(f) = K\,e^{-j2\pi fa} \qquad (5.9)$$

This function is already in polar form. The amplitude is the constant

$$|H(f)| = K$$

while the phase is

$$\theta_H(f) = -2\pi fa$$

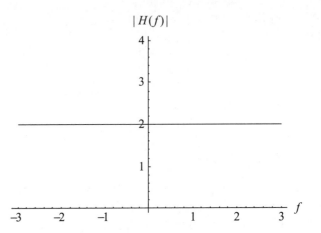

Fig. 5-4. A hypothetical amplitude spectrum for the case of distortionless transmission.

Summarizing, for a system with distortionless transmission the frequency response has two characteristics:

- The amplitude is constant for all frequencies and is equal to the gain constant.
- The phase varies linearly with frequency.

In Fig. 5-4, we show an example amplitude spectrum for a frequency response function with gain constant $K = 2$. In Fig. 5-5, we show a sample phase spectrum.

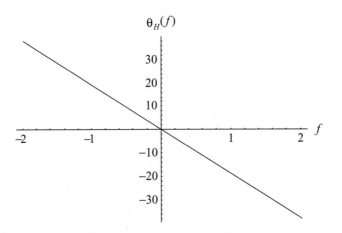

Fig. 5-5. The phase spectrum for $H(f)$ for a system with distortionless transmission and a time delay of 2 s.

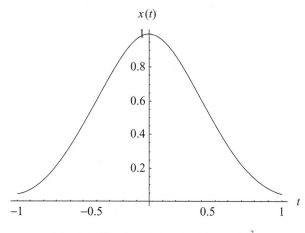

Fig. 5-6. The input signal $x(t) = e^{-\pi t^2}$.

EXAMPLE 5-3
A system is distortionless and described by a gain constant $K = 1/2$ and time delay $a = 3$ s. Describe the system output $y(t)$ given an input signal $x(t) = e^{-\pi t^2}$.

SOLUTION 5-3
The input signal is a Gaussian pulse, which we show in Fig. 5-6.

Using (5.7) together with the information given in the problem, the output signal is found to be

$$y(t) = \frac{1}{2}e^{-\pi(t-3)^2}$$

This is shown in Fig. 5-7. Notice that the shape has not changed, but the signal is shifted in time and the amplitude is decreased—characteristics of a distortionless system.

Filters

We now consider the operation of *filtering* or restricting the range of frequencies that are allowed through a device. A filter basically allows a specified range of frequencies to pass through and suppresses all other frequencies.

We call a given range of frequencies a *band*. The *passband* is the band of frequencies that are allowed through the filter. The *stopband* consists of those frequencies that are suppressed by the filter; i.e., they are not allowed to pass through. We begin by considering a set of *ideal* filters; we describe them by their

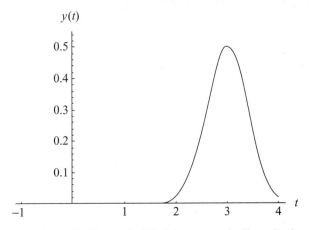

Fig. 5-7. The output signal for Example 5-3. The system is distortionless, so the shape of the signal is preserved; however the amplitude has been decreased and the signal is time-shifted.

frequency response function $H(f)$. The best way to visualize the function of a given filter type is to look at the amplitude spectrum of $H(f)$. We can generally group filters in four categories:

- *Low pass filters*: This type of filter allows low frequencies to pass through the filter to the output and rejects high frequencies.
- *High pass filters*: This is the reverse of the above; a high pass filter suppresses low frequencies and allows high frequencies to pass through.
- *Band pass*: A band pass filter allows a group of frequencies or a band to pass through the filter, and rejects frequencies that fall below the band or above the band.
- *Band stop*: The opposite of a band pass filter. Band stop filters reject frequencies that fall in a particular range.

The defining characteristic of an ideal filter is that the transition from the stop band to the pass band is *immediate*. You can think *unit step function* when you are trying to understand ideal filters. In contrast, a real filter is going to have a transition region between the stop and pass bands. Ideal filters are easier to understand and get the basic concepts across, so we start by considering the four types listed above.

IDEAL LOW PASS FILTER

An *ideal low pass filter* is defined by its cutoff frequency, which we label f_c. In a nutshell this type of filter allows frequencies to pass if $|f| < f_c$. The amplitude

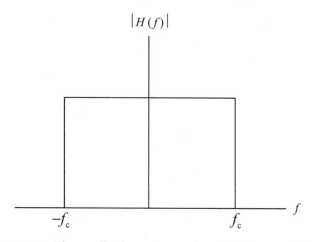

Fig. 5-8. The amplitude spectrum of an ideal low pass filter.

of an ideal low pass filter is described by

$$|H(f)| = \begin{cases} 1 & |f| < f_c \\ 0 & |f| > f_c \end{cases} \tag{5.10}$$

Remember the rectangular pulse? The amplitude spectrum of an ideal low pass filter has that same form, except now we are in the frequency domain. A plot of the amplitude spectrum is shown in Fig. 5-8.

The complete frequency response function for an ideal low pass filter is given by

$$H(f) = \begin{cases} e^{-j2\pi t_0 f} & |f| < f_c \\ 0 & |f| > f_c \end{cases} \tag{5.11}$$

The *bandwidth* of the filter is defined by its cutoff frequency f_c. Now, recalling that the rectangular pulse and the sinc function constitute a Fourier transform pair

$$rect\left(\frac{t}{T}\right) \rightleftharpoons T \operatorname{sinc}(Tf)$$

where $-T/2 \leq t \leq T/2$, we can use the duality property to find the inverse Fourier transform of a rectangle in the frequency domain. We have

$$rect\left(\frac{f}{2f_c}\right) \rightleftharpoons 2f_c \operatorname{sinc}(2f_c t)$$

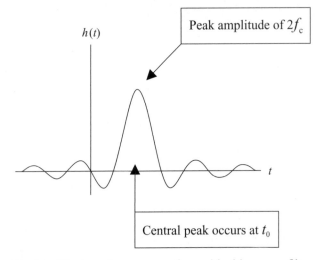

Fig. 5-9. The impulse response for an ideal low pass filter.

This gives us the impulse response function for the ideal low pass filter. Recalling the time shift property, (5.11) tells us that the sinc function will be shifted in time; i.e., $h(t) = 2f_c \, \text{sinc}\, [2f_c(t - t_0)]$. The peak of a sinc function is at the origin, and $\text{sinc}\,(\tau)$ has a peak amplitude of unity. Therefore, we see that the impulse response of an ideal low pass filter is centered about $t = t_0$ where it has a peak amplitude of $2f_c$. We show this in Fig. 5-9.

Given that $H(f) = e^{-j2\pi t_0 f}$, we see that the phase spectrum is a linearly decreasing function, but it is only nonzero between the values $\pm f_c$. We show a plot of the phase spectrum for an ideal low pass filter in Fig. 5-10.

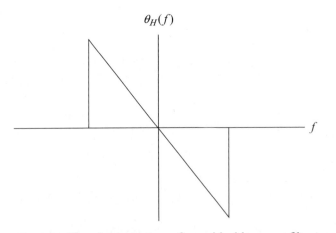

Fig. 5-10. The phase spectrum for an ideal low pass filter.

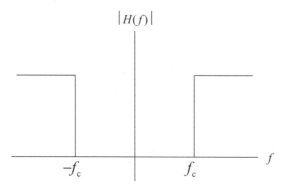

Fig. 5-11. The amplitude spectrum for the frequency response of an ideal high pass filter.

IDEAL HIGH PASS FILTER

Let us suppose that instead of allowing frequencies $|f| < f_c$ to pass we want to block those frequencies and allow frequencies $|f| > f_c$ to pass through. We accomplish this task with an *ideal high pass filter*. Formally, for the amplitude we write

$$|H(f)| = \begin{cases} 0 & |f| < f_c \\ 1 & |f| > f_c \end{cases} \qquad (5.12)$$

The amplitude spectrum is shown in Fig. 5-11.

In the same way that the amplitude spectrum is basically the opposite of that for the ideal low pass filter, the phase spectrum follows a similar pattern. Shown in Fig. 5-12, notice that the phase spectrum is zero inside the stop band frequency range but is linearly decaying everywhere else.

IDEAL BAND PASS FILTER

An ideal band pass filter is one that allows frequencies to pass if they fall between two specified frequencies; we will label f_1 and f_2 such that $f_1 < f_2$. In this case $|H(f)|$ is defined by

$$|H(f)| = \begin{cases} 1 & f_1 < |f| < f_2 \\ 0 & \text{otherwise} \end{cases} \qquad (5.13)$$

The amplitude spectrum is shown in Fig. 5-13.

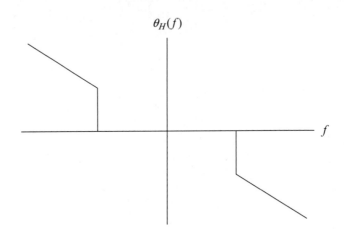

Fig. 5-12. A schematic representation of the phase spectrum for an ideal high pass filter.

IDEAL BAND STOP FILTER

An *ideal band stop filter* is one that rejects a band of frequencies. The amplitude spectrum is described by

$$|H(f)| = \begin{cases} 0 & f_1 < |f| < f_2 \\ 1 & \text{otherwise} \end{cases} \tag{5.14}$$

This is the opposite of the band pass filter. The amplitude spectrum is shown in Fig. 5-14. We show the phase spectrum in Fig. 5-15.

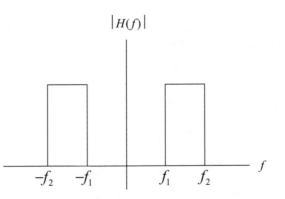

Fig. 5-13. The amplitude spectrum for a band pass filter.

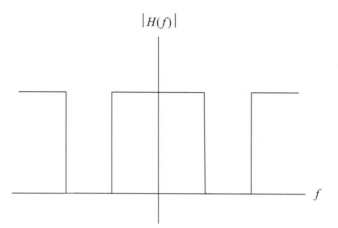

Fig. 5-14. An ideal band stop filter blocks out a range of frequencies that lie in a chosen band.

EXAMPLE 5-4

Consider the periodic function of Example 4-2, where $x(t) = t^2$ for $-1 < t < 1$ was extended on the real line. Suppose that this signal is passed through an ideal low pass filter described by

$$|H(\omega)| = \begin{cases} 1 & |\omega| < 3\pi \\ 0 & |\omega| > 3\pi \end{cases}$$

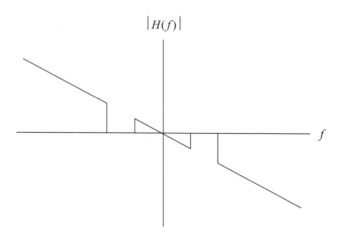

Fig. 5-15. The phase spectrum for an ideal band stop filter.

Find the output signal $y(t)$ for this filter. Consider what happens if the cutoff frequency is reduced to $\omega_c = 2\pi$.

SOLUTION 5-4

Notice that for this problem we are working with angular frequency. The cutoff frequency for the first part of the problem is given by $\omega_c = 3\pi$. In Example 4-2, we had found the Fourier series expansion of the input signal to be given by

$$x(t) = \frac{1}{3} - \frac{4}{\pi^2} \cos \pi t + \frac{1}{\pi^2} \cos 2\pi t - \frac{4}{9\pi^2} \cos 3\pi t + \cdots$$

The filter will pass all harmonic components with angular frequency $|\omega| < 3\pi$. The filter will reject all other harmonic components. Therefore the output signal is

$$y(t) = \frac{1}{3} - \frac{4}{\pi^2} \cos \pi t + \frac{1}{\pi^2} \cos 2\pi t$$

In Fig. 5-16, we show a plot of this output signal and compare it with the original input signal. The shape of the signal has been distorted somewhat by rejecting all but a few of the harmonic components, especially near the edges.

If we apply an even more restrictive filter with cutoff frequency $\omega_c = 2\pi$, then the output signal is

$$y(t) = \frac{1}{3} - \frac{4}{\pi^2} \cos \pi t$$

In Fig. 5-17, we see that this distorts the signal even further.

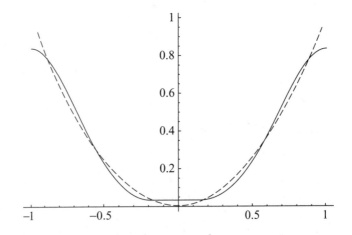

Fig. 5-16. Comparing $y(t) = \frac{1}{3} - \frac{4}{\pi^2} \cos \pi t + \frac{1}{\pi^2} \cos 2\pi t$ (solid line) to the original input signal, dashed line.

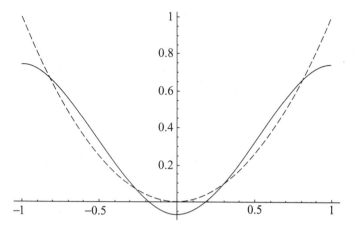

Fig. 5-17. The output signal is further degraded by rejecting one more harmonic component.

EXAMPLE 5-5
An ideal low pass filter has frequency response

$$|H(\omega)| = \begin{cases} 1 & |\omega| < 3 \\ 0 & |\omega| > 3 \end{cases}$$

Suppose that an input signal to this filter is $x(t) = 1/(1 + t^2)$. Find the ratio of the energy content of the output signal to that of the energy content of the input signal.

SOLUTION 5-5
In Fig. 5-18, we show the input signal.
   The energy content of this signal is

$$E_X = \int_{-\infty}^{\infty} |x(t)|^2 \, dt = \int_{-\infty}^{\infty} \left| \frac{1}{1 + t^2} \right|^2 \, dt$$

$$= \frac{1}{2} \left[\frac{t}{1 + t^2} + \arctan(t) \right] \Big|_{-\infty}^{\infty} = \frac{\pi}{2}$$

We can find the Fourier transform of the input signal from the Fourier transform pair

$$\frac{1}{a^2 + t^2} \rightleftharpoons e^{-a|\omega|}$$

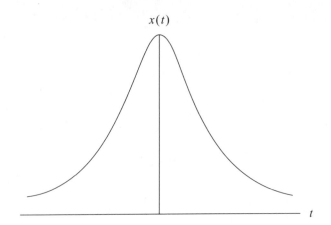

Fig. 5-18. The input signal $x(t) = 1/(1+t^2)$ passed through the filter of Example 5-5.

And so $X(\omega) = e^{-|\omega|}$. We can find the output signal in the frequency domain by using (5.1). This is particularly simple in this case given that we are working with an ideal low pass filter. We have

$$Y(f) = X(f)H(f) = \begin{cases} e^{-|\omega|} & |\omega| < 3 \\ 0 & \text{otherwise} \end{cases}$$

To find the energy content of the output signal, it is not necessary to find the form of the signal in the time domain and integrate; instead we call upon Parseval's identity. Using angular frequency this tells us that we can compute the energy content in the frequency domain since

$$\int_{-\infty}^{\infty} |x(t)|^2 \, dt = \frac{1}{2\pi} \int_{-\infty}^{\infty} |X(\omega)|^2 \, d\omega \tag{5.15}$$

The cutoff frequency of $\omega_c = 3$ rad/s makes this particularly simple, and so it is very convenient to find the energy content in the frequency domain for this example. We have

$$E_Y = \frac{1}{2\pi} \int_{-\infty}^{\infty} |Y(\omega)|^2 \, d\omega = \frac{1}{2\pi} \int_{-3}^{3} |Y(\omega)|^2 \, d\omega = \frac{1}{2\pi} \left(1 - e^{-6}\right)$$

The ratios of the energy content of the output signal to the input signal is then

$$\frac{E_Y}{E_X} = \frac{(1/2\pi)\left(1 - e^{-6}\right)}{(\pi/2)} \approx 0.10$$

So, we have found that the filter produces an output signal with only about 10% of the energy content of the input signal.

Causality and Filters

Ideal filters are not causal. This is because they have nonzero impulse response functions even when $t < 0$ (recall Fig. 5-9). A realistic filter must be causal. The criterion to determine whether a nonideal filter is causal is that $h(t) = 0$ for $t < 0$.

System Bandwidth

We close the chapter with a brief discussion of bandwidth for nonideal filters. A nonideal filter will have a transition region between the stop band and the pass band. As a result, the bandwidth is not as clear-cut as it is for an ideal filter. Let's quickly review the bandwidths of ideal filters.

The bandwidth of an ideal low pass filter is simply given by the cutoff frequency. If we denote the bandwidth by B, then in this case $B = \omega_c$. For an ideal band pass filter, the bandwidth is simply the difference between the two cutoff frequencies used to define the pass band. If we call these frequencies ω_1 and ω_2 with $\omega_1 < \omega_2$, then the bandwidth is given by $B = \omega_2 - \omega_1$. The midband frequency for a band pass filter is $\omega_0 = (1/2)(\omega_1 + \omega_2)$. If the bandwidth of a band pass filter is very small compared to the midband frequency, then we say that the filter is a *narrowband* filter.

For nonideal filters, due to the transition band it is not really possible to define bandwidth so simply. What is done is we appeal to a concept known as the 3-dB bandwidth. If ω_0 is the central or midpoint frequency, then we find the frequency at which

$$|H(\omega)| = \frac{|H(\omega_0)|}{\sqrt{2}}$$

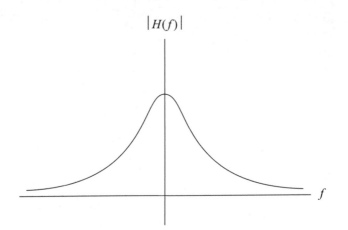

Fig. 5-19. An amplitude spectrum for a nonideal low pass filter.

If we call the frequency at which this condition is met ω_B, then the 3-dB bandwidth is given by

$$\omega_B - \omega_0 \tag{5.16}$$

For a low pass filter, we take $\omega_0 = 0$. As we have mentioned, a realistic filter has a transition band between the stop band and the pass band, rather than the immediate switch that we have seen with the ideal filters. For a low pass or band pass filter, the amplitude spectrum has a kind of bell shape. In Fig. 5-19, we see an example of a realistic low pass filter.

Quiz

1. Suppose that

$$H(f) = \begin{cases} -j & f > 0 \\ j & f < 0 \end{cases} = -j\,\mathrm{sgn}\,(f)$$

   What is the impulse response function $h(t)$?
2. Using the frequency response function of Problem 1, find the output $y(t)$ given an input signal $x(t) = \cos 2\pi f_0 t$.
3. Suppose that $dy/dt = x(t)$ and $x(t) = 4\delta(t-1)$. Find $y(t)$.
4. A system is described by $(dy/dt) + y(t) = x(t)$. Find the output when $x(t) = u(t)$.

5. A system is distortionless and has a time delay of $a = 1$ s. Plot the phase spectrum.

6. Consider an ideal low pass filter described by

$$|H(\omega)| = \begin{cases} 1 & |\omega| < 2\pi \\ 0 & |\omega| > 2\pi \end{cases}$$

Consider the input signal $x(t) = t, \ -1 < t < 1$. Describe the output signal $y(t)$ and plot it comparing to the input. The Fourier series expansion is given by

$$x(t) = \frac{1}{\pi} \sin \pi t - \frac{1}{2\pi} \sin 2\pi t + \frac{1}{3\pi} \sin 3\pi t - \frac{1}{4\pi} \sin 4\pi t + \cdots$$

For the next four problems, suppose that an ideal low pass filter is given by

$$|H(\omega)| = \begin{cases} 1 & |\omega| < \omega_c \\ 0 & |\omega| > \omega_c \end{cases}$$

7. Given an input signal $x(t) = e^{-t}u(t)$, find $X(\omega)$. What is the energy content of the input signal?

8. Find the form of the output signal in the frequency domain.

9. What is the energy content of the output signal?

10. Find the cutoff frequency such that the filter passes 50% of the energy content of the input signal.

CHAPTER 6

Energy Spectral Density and Correlation

In this chapter we begin by studying *correlation*. We will study two types of correlation, *cross-correlation*, which is the correlation between two signals, and *autocorrelation*, which is the cross-correlation of a signal with itself.

Next, we will consider the characterization of the energy and power content of signals in the frequency domain. When considering a frequency spectrum, we will be interested in determining the energy or power per bandwidth, leading to the notion of *spectral density*. Recalling from the first chapter that signals can be classified as energy signals or power signals, it will follow that we can calculate *energy spectral density* or *power spectral density* depending on which type of signal we are working with.

Cross-Correlation

Cross-correlation is a way to measure the similarity between two energy signals. For example, we might wish to measure the properties of an unknown signal by comparing it to a known signal. The cross-correlation between two signals $x_1(t)$ and $x_2(t)$ is a function of time that has a form that is somewhat analogous to convolution. Specifically, we compare $x_1(t)$ to a time-delayed version of $x_2(t)$. The cross-correlation function $R_{12}(\tau)$ is given by

$$R_{12}(\tau) = \int_{-\infty}^{\infty} x_1(t)x_2(t - \tau)\,dt \tag{6.1}$$

We can think of the parameter τ as a time-lag, which acts as a scanning parameter. That is, we are searching or scanning one unknown signal to see if it contains features that are present in the known signal. If the two signals do correlate in some way, then (6.1) will have a finite but nonzero value over some range of τ. To see why this is the case think of orthogonal functions. Two functions $f(t)$ and $g(t)$ are *orthogonal* provided that the integral of their product vanishes:

$$\int_{-\infty}^{\infty} f(t)g(t)\infty\,dt = 0$$

For example, suppose that $f(t) = \cos t$ and $g(t) = \sin t$. We plot these functions in Fig. 6-1, showing $\cos t$ as the solid line and $\sin t$ as the dashed line. In a

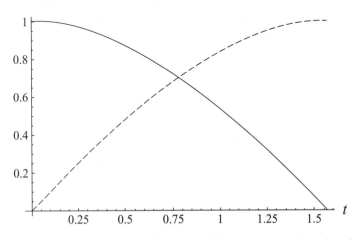

Fig. 6-1. The functions $f(t) = \cos t$ and $g(t) = \sin t$ are orthogonal, and so their integral vanishes.

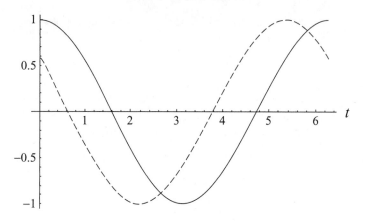

Fig. 6-2. A plot of $f(t) = \cos t$ (solid line) and $g(t) = \sin(t - 6\pi/5)$ (dashed line).

sense, these functions are opposites. Consider how these functions behave over the range $0 \le t \le \pi/2$. Notice that when $\cos t$ is *decreasing*, $\sin t$ is *increasing*. When $\cos t$ is at its maximum, $\sin t$ is at its minimum, and vice versa. These functions do not seem very similar. If we integrate over a full period, this idea gets some validation, since the integral vanishes:

$$\int_0^{2\pi} \cos t \, \sin t \, dt = \left. \frac{\sin^2 t}{2} \right|_0^{2\pi} = \frac{\sin^2 2\pi}{2} - \frac{\sin^2 0}{2} = 0 - 0 = 0$$

Now let's consider a time-shifted sine function. This time we continue to take $f(t) = \cos t$ but instead set $g(t) = \sin(t - 6\pi/5)$. We show a plot of these two functions in Fig. 6-2, again plotting $f(t) = \cos t$ with a solid line.

This time the behavior of the two functions seems to be similar, or *correlated*. While $g(t) = \sin(t - 6\pi/5)$ can be viewed as leading $f(t) = \cos t$ a bit in time, we see that the two signals have basically the same shape and behavior. This time the integral of their product is nonzero:

$$\int_0^{2\pi} \cos(t) \, \sin(t - 6\pi/5) \, dt = \frac{\pi}{4}\sqrt{10 - 2\sqrt{5}}$$

Returning to the cross-correlation function, notice from (6.1) that if we exchange the roles of the two signals we obtain

$$R_{21}(\tau) = \int_{-\infty}^{\infty} x_2(t) x_1(t - \tau) dt$$

From this we see that generally speaking, the cross-correlation is not commutative; i.e.,

$$R_{12}(\tau) \neq R_{21}(\tau)$$

In the frequency domain, cross-correlation corresponds to a multiplication. This is to be expected due to the similarity of (6.1) with convolution. In this case, however, we must take the complex conjugate of the second function in the frequency domain. Specifically we have the following Fourier transform pair:

$$R_{12}(\tau) \rightleftharpoons X_1(f) X_2(f) \tag{6.2}$$

Autocorrelation

Autocorrelation is the cross-correlation of a signal with itself. Again we time-delay the signal and scan for patterns. In short we are finding repeating patterns in a signal. The formula used for autocorrelation is

$$R_{11}(\tau) = \int_{-\infty}^{\infty} x_1(t)x_1(t - \tau) \, dt \tag{6.3}$$

The normalized energy content of a signal is related to the autocorrelation function by setting $\tau = 0$ in (6.3), giving

$$R_{11}(0) = \int_{-\infty}^{\infty} |x_1(t)|^2 \, dt = E \tag{6.4}$$

EXAMPLE 6-1
Compute the autocorrelation function for $x(t) = e^{-at}u(t)$. Assume that $\tau > 0$.

SOLUTION 6-1
Using (6.3) we set the integral up as

$$R_{11}(\tau) = \int_{-\infty}^{\infty} xt(t)x(t - \tau) \, dt = \int_{-\infty}^{\infty} \left[e^{-at}u(t)\right] e^{-a(t-\tau)}u(t - \tau) \, dt$$

To simplify the integral, recall that

$$u(t) = \begin{cases} 1 & t \geq 0 \\ 0 & t < 0 \end{cases}$$

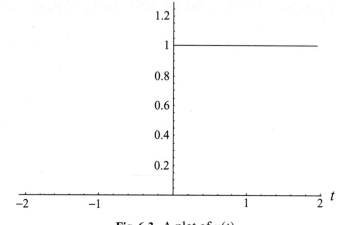

Fig. 6-3. A plot of $u(t)$.

This function is plotted in Fig. 6-3.

Since $\tau > 0$, the other unit step function in the integral is

$$u(t - \tau) = \begin{cases} 1 & t \geq \tau \\ 0 & t < \tau \end{cases}$$

As an example, consider $\tau = 2$, shown in Fig. 6-4.

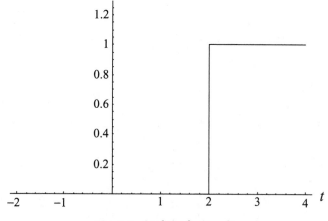

Fig. 6-4. A plot of $u(t - 2)$.

Looking at the plots, notice that the product is zero for $t < \tau$. So, the product of these two functions is defined by

$$u(t)u(t - \tau) = \begin{cases} 1 & t \geq \tau \\ 0 & t < \tau \end{cases}$$

With this in mind, the autocorrelation function becomes

$$R_{11}(\tau) = \int_{\tau}^{\infty} \left(e^{-at}\right) e^{-a(t-\tau)} \, dt = e^{a\tau} \int_{\tau}^{\infty} e^{-2at} \, dt$$

$$= -\frac{e^{a\tau}}{2a} e^{-2at} \Big|_{\tau}^{\infty} = \frac{e^{a\tau}}{2a} e^{-2a\tau} = \frac{e^{-a\tau}}{2a}$$

AUTOCORRELATION IN POWER SIGNALS

To find the autocorrelation of a power signal we compute the time-average of the autocorrelation function. That is, we compute

$$\bar{R}_{11}(\tau) = \lim_{T \to \infty} \frac{1}{T} \int_{-T/2}^{T/2} x_1(t) x_1(t - \tau) \, dt \tag{6.5}$$

In the case of a periodic signal, we compute the time-average autocorrelation function over the period T_0

$$\bar{R}_{11}(\tau) = \frac{1}{T_0} \int_{-T_0/2}^{T_0/2} x_1(t) x_1(t - \tau) \, dt \tag{6.6}$$

Setting $\tau = 0$ in (6.5) gives the average power in the signal

$$\bar{R}_{11}(0) = \lim_{T \to \infty} \frac{1}{T} \int_{-T/2}^{T/2} |x_1(t)|^2 \, dt = P \tag{6.7}$$

EXAMPLE 6-2

Find the time-average autocorrelation function for the signal $x(t) = A \cos \omega t$.

SOLUTION 6-2

Since $x(t) = A \cos \omega t$ is periodic we use (6.6). We call the period T so that

$$\omega = \frac{2\pi}{T}$$

Setting up the autocorrelation function (6.6) gives

$$\bar{R}_{11}(\tau) = \frac{1}{T} \int_{-T/2}^{T/2} x(t)x(t - \tau) \, dt$$

$$= \frac{1}{T} \int_{-T/2}^{T/2} A \cos(\omega t) \, A \cos[\omega(t - \tau)] \, dt$$

We expand $\cos[\omega(t - \tau)]$ using a trig identity

$$\cos[\omega(t - \tau)] = \cos \omega t \, \cos \omega \tau + \sin \omega t \, \sin \omega \tau$$

The autocorrelation function can then be written as

$$\bar{R}_{11}(\tau) = \frac{A^2}{T} \int_{-T/2}^{T/2} \cos(\omega t) \, \cos[\omega(t - \tau)] \, dt$$

$$= \frac{A^2}{T} \int_{-T/2}^{T/2} \cos(\omega t) [\cos \omega t \, \cos \omega \tau + \sin \omega t \, \sin \omega \tau] \, dt$$

And so we have two integrals to compute

$$\bar{R}_{11}(\tau) = \frac{A^2}{T} \int_{-T/2}^{T/2} \cos^2(\omega t) \, \cos \omega \tau \, dt + \frac{A^2}{T} \int_{-T/2}^{T/2} \cos(\omega t) \sin \omega t \, \sin \omega \tau \, dt$$

Let's look at the second integral first. The integration is over t and so we can write this as

$$\frac{A^2}{T} \int_{-T/2}^{T/2} \cos(\omega t) \sin \omega t \, \sin \omega \tau \, dt = \frac{A^2 \sin \omega \tau}{T} \int_{-T/2}^{T/2} \cos(\omega t) \sin \omega t \, dt$$

Now we set $u = \sin \omega t$ and so $du = \omega \cos \omega t \, dt$. Now notice that at the upper limit $t = T/2$ we have

$$u = \sin[\omega(T/2)] = \sin\left[\tfrac{2\pi}{T} (T/2)\right] = \sin \pi = 0$$

We obtain the same result at the lower limit and so the integral vanishes.
Turning our attention to the first integral, we write

$$\bar{R}_{11}(\tau) = \frac{A^2}{T} \int_{-T/2}^{T/2} \cos^2(\omega t) \cos \omega \tau \, dt$$

$$= \frac{A^2 \cos \omega \tau}{T} \int_{-T/2}^{T/2} \left(\frac{1 + \cos 2\omega t}{2} \right) dt$$

Now since

$$\int_{-T/2}^{T/2} \cos 2\omega t \, dt = \frac{1}{2\omega} \sin 2\omega t \bigg|_{-T/2}^{T/2} = \frac{1}{2\omega} [\sin 2\pi - \sin(-2\pi)] = 0$$

We are left with

$$\bar{R}_{11}(\tau) = \frac{A^2 \cos \omega \tau}{2T} \int_{-T/2}^{T/2} dt = \frac{A^2 \cos \omega \tau}{2T} \left(\frac{T}{2} + \frac{T}{2} \right) = \frac{A^2 \cos \omega \tau}{2}$$

Energy Spectral Density

The *energy spectral density* of an energy signal is found by taking the Fourier transform of the autocorrelation function. We denote the energy spectral density by $S_{11}(\omega)$. The formal definition is given by

$$S_{11}(\omega) = \int_{-\infty}^{\infty} R_{11}(\tau) e^{-j\omega\tau} \, d\tau \qquad (6.8)$$

If the signal $x(t)$ is a real-valued signal then

$$S_{11}(\omega) = |X(\omega)|^2 \qquad (6.9)$$

In particular, using (6.4) we can compute the energy in a signal as

$$E = R_{11}(0) = \int_{-\infty}^{\infty} |x_1(t)|^2 \, dt = \frac{1}{2\pi} \int_{-\infty}^{\infty} |X(\omega)|^2 \, d\omega \qquad (6.10)$$

Looking at (6.9), we see that

$$E = \frac{1}{2\pi} \int_{-\infty}^{\infty} |X(\omega)|^2 \, d\omega = \frac{1}{2\pi} \int_{-\infty}^{\infty} S_{11}(\omega) \, d\omega \qquad (6.11)$$

This is why we call $S_{11}(\omega)$ the energy spectral density. If we integrate it over the entire spectrum, we obtain the energy content of the signal.

EXAMPLE 6-3
Find the energy content of the signal used in Example 6-1 by considering the energy spectral density.

SOLUTION 6-3
We found that

$$R_{11}(\tau) = \frac{e^{-a\tau}}{2a} u(\tau)$$

We have included the unit step function because in that example it had been specified that $\tau \geq 0$. The Fourier transform of this function is

$$\text{FT}[R_{11}(\tau)] = \text{FT}\left[\frac{e^{-a\tau}}{2a} u(\tau)\right] = \frac{1}{a + j\omega}$$

Using (6.9), the energy spectral density is given by

$$S_{11}(\omega) = \left|\frac{1}{a + j\omega}\right|^2 = \left(\frac{1}{a + j\omega}\right)\left(\frac{1}{a - j\omega}\right) = \frac{1}{a^2 + \omega^2}$$

Now recall that

$$\int \frac{1}{1 + x^2} dx = \arctan x$$

Evaluating this at the limits $x \to \pm\infty$ we obtain the result

$$\int_{-\infty}^{\infty} \frac{1}{1 + x^2} dx = \pi$$

So we have

$$
\int_{-\infty}^{\infty} S_{11}(\omega)\, d\omega = \int_{-\infty}^{\infty} \frac{1}{a^2 + \omega^2}\, d\omega = \int_{-\infty}^{\infty} \frac{a^2}{1 + (\omega/a)^2}\, d\omega
$$

$$
= a \int_{-\infty}^{\infty} \frac{1}{1 + x^2}\, dx = a\pi
$$

Using (6.11), we find that the total energy content of the signal is

$$
E = \frac{1}{2\pi} \int_{-\infty}^{\infty} S_{11}(\omega)\, d\omega = \frac{a\pi}{2\pi} = \frac{a}{2}
$$

Power Spectral Density

We close the chapter with a look at *power spectral density*. This is defined in an analogous manner to the energy spectral density, in the same way that we defined the time-average autocorrelation function. The power spectral density is denoted by $\bar{S}_{11}(\omega)$ and is computed as the Fourier transform of the time-average autocorrelation:

$$
\bar{S}_{11}(\omega) = \text{FT}\left[\bar{R}_{11}(\tau)\right] = \int_{-\infty}^{\infty} \bar{R}_{11}(\tau)\, e^{-j\omega\tau}\, d\tau \tag{6.12}
$$

We can find the power in a signal by calling on (6.7). With $\tau = 0$ we have

$$
P = \bar{R}_{11}(0) = \frac{1}{2\pi} \int_{-\infty}^{\infty} \bar{S}_{11}(\omega)\, d\omega \tag{6.13}
$$

EXAMPLE 6-4
Find the average power in the signal $x(t) = A\,\cos \omega_0 t$ and check (6.13).

SOLUTION 6-4
In Example 6-2 we found that

$$
\bar{R}_{11}(\tau) = \frac{A^2\,\cos \omega_0 \tau}{2}
$$

Setting $\tau = 0$, we find that

$$P = \frac{A^2}{2}$$

The power spectral density is the Fourier transform of the time-average autocorrelation function. We find that

$$\bar{S}_{11}(\omega) = \text{FT}\left[\bar{R}_{11}(\tau)\right]$$

$$= \text{FT}\left[\frac{A^2 \cos \omega_0 \tau}{2}\right]$$

$$= \frac{A^2}{2}\pi \left[\delta(\omega - \omega_0) + \delta(\omega + \omega_0)\right]$$

Integrating we have

$$P = \frac{1}{2\pi}\int_{-\infty}^{\infty} \bar{S}_{11}(\omega)\,d\omega$$

$$= \frac{1}{2\pi}\left(\frac{A^2}{2}\pi\right)\left[\int_{-\infty}^{\infty}\delta(\omega - \omega_0)\,d\omega + \int_{-\infty}^{\infty}\delta(\omega + \omega_0)\,d\omega\right]$$

Using the properties of the unit impulse function, we find

$$P = \frac{1}{2\pi}\left(\frac{A^2}{2}\pi\right)\left[\int_{-\infty}^{\infty}\delta(\omega - \omega_0)\,d\omega + \int_{-\infty}^{\infty}\delta(\omega + \omega_0)\,d\omega\right]$$

$$= \frac{1}{2\pi}\left(\frac{A^2}{2}\pi\right)(1 + 1) = \frac{A^2}{2}$$

Quiz

1. The autocorrelation function for $x(t) = e^{-at}u(t)$ and $a > 0$ with τ arbitrary is
   (a) $(1/2a)\,e^{-a|\tau|}$
   (b) $(1/2a)\,e^{-a\tau}$
   (c) $(1/2a)\,e^{a\tau}$
   (d) $-(1/2a)\,e^{-a\tau}$

2. The time-average autocorrelation function of $x(t) = A \sin \omega t$ is
   (a) $(A^2 \sin \omega \tau / 2)$
   (b) 0
   (c) $(A^2 \cos \omega \tau / 2)$
   (d) $(A^2 \sin \omega \tau \cos \omega \tau / 2)$

3. By using the power spectral density, the power content of the signal $x(t) = A \sin \omega t$ is given by
   (a) $P = 0$
   (b) $P = A/2$
   (c) $P = A^2/2$
   (d) $P = A^2$

4. For a real-valued signal $x(t)$, we can define the energy spectral density as
   (a) $\bar{S}_{11}(\omega) = |X(\omega)|^2$
   (b) $S_{11}(\omega) = |X(\omega)|^2 / 2\pi$
   (c) $S_{11}(\omega) = |X(\omega)|$
   (d) $S_{11}(\omega) = |X(\omega)|^2$

5. The cross-correlation function is part of the Fourier transform pair
   (a) $R_{12}(\tau) \rightleftharpoons X_1(f)X_2(f)$
   (b) $R_{12}(\tau) \rightleftharpoons X_1(f)X_2(-f)$
   (c) $R_{12}(\tau) \rightleftharpoons X_1(f)X_2^*(f)$
   (d) $R_{12}(\tau) \rightleftharpoons X_1(f)X_2^*(-f)$

CHAPTER 7

Discrete Fourier Transform

Consider a discrete time signal $x[n]$ with fundamental period N_0 as

$$x[n] = x[n + N_0] \tag{7.1}$$

The fundamental frequency Ω_0 is related to the fundamental period via the relationship

$$\Omega_0 = \frac{2\pi}{N_0} \tag{7.2}$$

The Fourier series representation of (7.1) is given by

$$x[n] = \sum_{k=0}^{N_0-1} c_k \, e^{jk\Omega_0 n} \tag{7.3}$$

We call the C_k *Fourier* or *spectral coefficients*, and they are given by

$$c_k = \frac{1}{N_0} \sum_{n=0}^{N_0-1} x[n] e^{-jk\Omega_0 n} \tag{7.4}$$

Often we let the sum run over any N_0 consecutive values of k. The notation used to indicate this for (7.3) is

$$x[n] = \sum_{k=\langle N_0 \rangle} c_k e^{jk\Omega_0 n} \tag{7.5}$$

This equation (7.5) is sometimes known as the *synthesis equation*. Using the same notation, we can write (7.4) in the following way:

$$c_k = \frac{1}{N_0} \sum_{n=\langle N_0 \rangle} x[n] e^{-jk\Omega_0 n} \tag{7.6}$$

The defining equation for the spectral coefficients (7.6) is sometimes called the *analysis equation*. We say that the spectral coefficients and the sequence $x[n]$ constitute a Fourier series pair and denote this by writing

$$x[n] \quad \rightleftharpoons \quad C_k$$

The average value of $x[n]$ over a period is given by c_0. In particular, c_0 is

$$c_0 = \frac{1}{N_0} \sum_{n=\langle N_0 \rangle} x[n] \tag{7.7}$$

EXAMPLE 7-1
Find the spectral coefficients for the discrete time signal shown in Fig. 7-1.

SOLUTION 7-1
We can see from Fig. 7-1 that the fundamental period of this signal is

$$N_0 = 4$$

We can write the sequence as

$$x[n] = \{4, \ 1, \ 2, \ 3\} \tag{7.8}$$

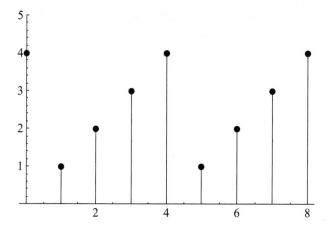

Fig. 7-1. The discrete time signal used in Example 7-1.

The fundamental frequency, using (7.2), is

$$\Omega_0 = \frac{2\pi}{N_0} = \frac{2\pi}{4} = \frac{\pi}{2}$$

Before we proceed, we compute a few values of $e^{-j\theta} = \cos\theta - j\sin\theta$ that will be useful in this problem. These are shown in Table 7-1.

We calculate each of the spectral coefficients using (7.4). Starting with the $k = 0$ case, we obtain

$$c_0 = \frac{1}{4}\sum_{n=0}^{3} x[n] = \frac{1}{4}(4 + 1 + 2 + 3) = \frac{1}{4}(10) = \frac{5}{2} \tag{7.9}$$

Table 7-1. Useful results using Euler's identity for Example 7-1.

Angle	Value
$\theta = \pi/2$	$e^{-j\pi/2} = \cos(\pi/2) - j\sin(\pi/2) = -j$
$\theta = \pi$	$e^{-j\pi} = \cos(\pi) - j\sin(\pi) = -1$
$\theta = 3\pi/2$	$e^{-j3\pi/2} = \cos(3\pi/2) - j\sin(3\pi/2) = j$
$\theta = 2\pi$	$e^{-j2\pi} = \cos(2\pi) - j\sin(2\pi) = 1$
$\theta = 3\pi$	$e^{-j3\pi} = \cos(3\pi) - j\sin(3\pi) = -1$
$\theta = 9\pi/2$	$e^{-j9\pi/2} = \cos(9\pi/2) - j\sin(9\pi/2) = -j$

Next, we compute c_1 as

$$c_1 = \frac{1}{4} \sum_{n=0}^{3} x[n] e^{-j\frac{\pi}{2}n} = \frac{1}{4}\left(4 + e^{-j\frac{\pi}{2}} + 2 e^{-j\pi} + 3 e^{-j\frac{3\pi}{2}}\right)$$

$$= \frac{1}{4}(4 - j - 2 + 3j) = \frac{1}{2}(1 + j) \qquad (7.10)$$

Moving to $k = 2$, we find

$$c_2 = \frac{1}{4} \sum_{n=0}^{3} x[n] e^{-j\pi n} = \frac{1}{4}(4 + e^{-j\pi} + 2 e^{-j2\pi} + 3 e^{-j3\pi})$$

$$= \frac{1}{4}(4 - 1 + 2 - 3) = \frac{1}{2}$$

The last case we need to consider is $k = 3$ to find

$$c_3 = \frac{1}{4} \sum_{n=0}^{3} x[n] e^{-j\frac{3\pi}{2}n} = \frac{1}{4}\left(4 + e^{-j\frac{3\pi}{2}} + 2 e^{-j3\pi} + 3 e^{-j\frac{9\pi}{2}}\right)$$

$$= \frac{1}{4}(4 + j - 2 - 3j) = \frac{1}{2}(1 - j)$$

Let's check the results using (7.3) for a couple of cases. For $n = 0$, using (7.9), we have

$$x[0] = \sum_{k=0}^{3} c_k = \frac{5}{2} + \frac{1}{2}(1 + j) + \frac{1}{2} + \frac{1}{2}(1 - j)$$

$$= \frac{5}{2} + \frac{1}{2} + \frac{1}{2} + \frac{1}{2} = \frac{8}{2} = 4$$

And for $n = 1$, we have

$$x[1] = \sum_{k=0}^{3} c_k e^{jk\pi/2}$$

$$= 5/2 + 1/2\,(1 + j)\, e^{j\pi/2} + (1/2)\, e^{j\pi} + 1/2\,(1 - j)\, e^{j3\pi/2}$$

$$= 5/2 + 1/2\,(1 + j)(j) + (1/2)(-1) + 1/2\,(1 - j)(-j)$$

$$= 5/2 + (j/2) - 1/2 - 1/2 - (j/2) - 1/2$$

$$= 1$$

EXAMPLE 7-2

Compute c_3 for the sequence given in Example 7-1, this time using (7.6) and starting at $n = 2$.

SOLUTION 7-2

Using the fundamental period $N_0 = 4$, starting at $n = 2$, the summation used in (7.6) will run over $n = 2, 3, 4, 5$. We have

$$c_3 = \frac{1}{4} \sum_{n=2}^{5} x[n] e^{-j\frac{3\pi}{2}n}$$

$$= \frac{1}{4} \left(x[2] e^{-j3\pi} + x[3] e^{-j\frac{9\pi}{2}} + x[4] e^{-j6\pi} + x[5] e^{-j\frac{15\pi}{2}} \right)$$

$$= \frac{1}{4} \left(2 e^{-j3\pi} + 3 e^{-j\frac{9\pi}{2}} + 4 e^{-j6\pi} + e^{-j\frac{15\pi}{2}} \right)$$

where the values of the $x[n]$ were obtained from Fig. 7-1. From Table 7-1, we see that

$$e^{-j3\pi} = -1$$

$$e^{-j\frac{9\pi}{2}} = -j$$

Recalling that $e^{j\theta}$ is periodic with period 2π, we determine that the other two exponentials in the above sum are

$$e^{-j6\pi} = e^{-j2\pi} = +1$$

$$e^{-j\frac{15\pi}{2}} = e^{-j\frac{3\pi}{2}} = +j$$

and so we have

$$c_3 = \frac{1}{4} \left(2 e^{-j3\pi} + 3 e^{-j\frac{9\pi}{2}} + 4 e^{-j6\pi} + e^{-j\frac{15\pi}{2}} \right)$$

$$= \frac{1}{4} (-2 - 3j + 4 + j) = \frac{1}{2} (1 - j)$$

As expected, this is the same result obtained using the sum in Example 7-1.

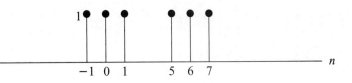

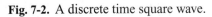

Fig. 7-2. A discrete time square wave.

EXAMPLE 7-3
Find the spectral coefficients for the discrete time square wave shown in Fig. 7-2.

SOLUTION 7-3
The fundamental period is $N_0 = 6$ and the fundamental frequency is

$$\Omega_0 = \frac{2\pi}{N_0} = \frac{2\pi}{6} = \frac{\pi}{3}$$

The coefficients are given by

$$c_{-1} = \frac{1}{6} \sum_{n=-1}^{1} e^{j\frac{\pi}{3}n} = \frac{1}{6}\left(e^{-j\frac{\pi}{3}} + 1 + e^{j\frac{\pi}{3}}\right)$$

$$= \frac{1}{6}\left(1 + 2\cos\frac{\pi}{3}\right) = \frac{1}{3}\cos^2\left(\frac{\pi}{6}\right) = \frac{1}{4}$$

The average value of the signal over one period is

$$c_0 = \frac{1}{6} \sum_{n=-1}^{1} x[n] = \frac{1}{6}(1 + 1 + 1) = \frac{1}{2}$$

And finally, we find that

$$c_1 = \frac{1}{6} \sum_{n=-1}^{1} e^{-j\frac{\pi}{3}n} = \frac{1}{6}\left(e^{j\frac{\pi}{3}} + 1 + e^{-j\frac{\pi}{3}}\right)$$

$$= \frac{1}{6}\left(1 + 2\cos\frac{\pi}{3}\right) = \frac{1}{3}\cos^2\left(\frac{\pi}{6}\right) = \frac{1}{4}$$

Some Properties of Fourier Series for Periodic Discrete Signals

So far we have been considering *periodic* discrete time signals, meaning that

$$x[n] = x[n + N_0]$$

for fundamental period N_0. It follows from (7.5) that the spectral coefficients c_k are also periodic with period N_0; that is,

$$c_k = c_{k+N_0} \qquad (7.11)$$

Using (7.6), which we restate here for convenience

$$c_k = \frac{1}{N_0} \sum_{n=\langle N_0 \rangle} x[n] e^{-jk\Omega_0 n}$$

we can view the members of the discrete time sequence as Fourier coefficients of the c_k. We can see this with some minor manipulation of this expression. First, we bring the constant $1/N_0$ inside the sum as

$$c_k = \sum_{n=\langle N_0 \rangle} \frac{x[n]}{N_0} e^{-jk\Omega_0 n}$$

Now we make a simple notational change and write $c_k = c[k]$ to give

$$c[k] = \sum_{n=\langle N_0 \rangle} \frac{x[n]}{N_0} e^{-jk\Omega_0 n}$$

Now let's change variables. First, we let $m = -n$ so that the above expression becomes

$$c[k] = \sum_{m=\langle N_0 \rangle} \frac{x[-m]}{N_0} e^{jk\Omega_0 m}$$

The variables used in this expression can be given new names at our convenience. So now we let $k \to n$ and $m \to k$ and obtain

$$c[n] = \sum_{k=\langle N_0 \rangle} \frac{x[-k]}{N_0} e^{jk\Omega_0 n} \tag{7.12}$$

This is just a discrete Fourier series representation for the $c[n]$. The relation between the sequences $x[n]$ and $c[n]$ which are related via Fourier series is known as the *duality property*. The duality property tells us that if $x[n]$ and $c[k]$ form a Fourier series pair

$$x[n] \rightleftharpoons c[k]$$

then we also have the Fourier series pair

$$c[n] \rightleftharpoons \frac{x[-k]}{N_0}$$

Finally, we list *Parseval's theorem* for discrete Fourier series as

$$\frac{1}{N_0} \sum_{n=\langle N_0 \rangle} |x[n]|^2 = \sum_{k=\langle N_0 \rangle} |c[k]|^2 \tag{7.13}$$

This tells us that we can find the average power of a discrete time signal by summing the squared amplitudes of its harmonic components.

EXAMPLE 7-4
Demonstrate Parseval's theorem for the signal in Example 7-1.

SOLUTION 7-4
Considering $x[n]$, we have

$$\frac{1}{N_0} \sum_{n=0}^{3} |x[n]|^2 = \frac{1}{4}\left(4^2 + 1^2 + 2^2 + 3^2\right) = \frac{1}{4}(16 + 1 + 4 + 9) = \frac{30}{4}$$

Now for the Fourier coefficients, we have

$$\sum_{k=0}^{3} |c[k]|^2 = \left[\left(\frac{5}{2}\right)^2 + \left|\frac{1}{2}(1+j)\right|^2 + \left(\frac{1}{2}\right)^2 + \left|\frac{1}{2}(1-j)\right|^2\right]$$

Now for a complex number z, we have $|z|^2 = zz^*$ and so

$$\left| \frac{1}{2}(1+j) \right|^2 = \frac{1}{4}(1+j)(1-j) = \frac{1}{4}(1+1) = \frac{1}{2}$$

And hence, the sum is

$$\sum_{k=0}^{3} |c[k]|^2 = \left[\frac{25}{4} + \frac{1}{2} + \frac{1}{4} + \frac{1}{2} \right] = \frac{30}{4}$$

verifying Parseval's theorem.

The Fourier Transform of a Discrete Time Signal

Consider an arbitrary nonperiodic discrete time signal $x[n]$. The *Fourier transform* of $x[n]$ is given by

$$X(\Omega) = \sum_{n=-\infty}^{\infty} x[n] e^{-j\Omega n} \qquad (7.14)$$

The Fourier transform defined by (7.14) is periodic in 2π, meaning that $X(\Omega) = X(\Omega + 2\pi)$. Moreover, the product $X(\Omega)\, e^{j\Omega n}$ is periodic with period 2π. The inverse Fourier transform is computed by integrating over an interval of length 2π as

$$x[n] = \frac{1}{2\pi} \int_{2\pi} X(\Omega)\, e^{j\Omega n}\, d\Omega \qquad (7.15)$$

It is easy to see, from the definition, that the Fourier transform of a discrete time signal is linear; i.e.,

$$ax_1[n] + bx_2[n] \rightleftharpoons aX_1(\Omega) + bX_2(\Omega) \qquad (7.16)$$

Time shifting by n_0 results in the Fourier transform pair

$$x[n - n_0] \rightleftharpoons e^{-j\Omega n_0} X(\Omega) \qquad (7.17)$$

while a frequency shift by Ω_0 gives

$$e^{j\Omega_0 n} x[n] \rightleftharpoons X(\Omega - \Omega_0) \tag{7.18}$$

Using the time-shifting property, we see that a difference equation $x[n] - x[n-1]$ gives us the Fourier transform pair

$$x[n] - x[n-1] \rightleftharpoons \left(1 - e^{-j\Omega}\right) X(\Omega) \tag{7.19}$$

The *accumulation property* is defined by the Fourier transform pair

$$\sum_{k=-\infty}^{n} x[k] \rightleftharpoons \pi X(0) \delta(\Omega) + \frac{1}{1 - e^{-j\Omega}} X(\Omega) \tag{7.20}$$

where we take $|\Omega| \leq \pi$.

EXAMPLE 7-5
Find the Fourier transform of the discrete time signal given by

$$x[n] = r^n u[n] \quad |r| < 1$$

SOLUTION 7-5
Using (7.14) we can write the Fourier transform of this signal as

$$X(\Omega) = \sum_{n=-\infty}^{\infty} r^n u[n] e^{-j\Omega n} = \sum_{n=0}^{\infty} r^n e^{-j\Omega n} = \sum_{n=0}^{\infty} \left(r e^{-j\Omega}\right)^n$$

Now since $|r| < 1$, this is a geometric series and we can write

$$X(\Omega) = \sum_{n=0}^{\infty} \left(r e^{-j\Omega}\right)^n = \frac{1}{1 - r e^{-j\Omega}}$$

Now let's multiply and divide by $1 - r e^{j\Omega}$ to write this in a more convenient form as

$$X(\Omega) = \frac{1}{1 - re^{-j\Omega}} \left(\frac{1 - re^{j\Omega}}{1 - re^{j\Omega}}\right)$$

$$= \frac{1 - re^{j\Omega}}{1 - re^{-j\Omega} - re^{j\Omega} + r^2} = \frac{1 - re^{j\Omega}}{1 - 2r \cos\Omega + r^2}$$

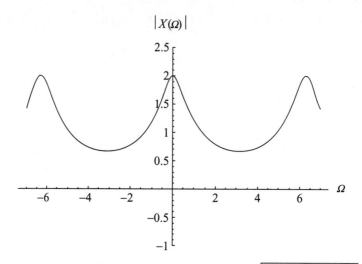

Fig. 7-3. The amplitude spectrum $|X(\Omega)| = 1/\sqrt{1 - 2r \cos \Omega + r^2}$.

With the Fourier transform expressed in this way, we see that the amplitude spectrum is given by

$$|X(\Omega)| = \frac{1}{\sqrt{1 - 2r \cos \Omega + r^2}}$$

The amplitude spectrum is shown in Fig. 7-3 where we have set $r = 1/2$. Notice that at $\Omega = 0$,

$$|X(\Omega)| = \frac{1}{\sqrt{1 - 2r + r^2}} = \frac{1}{\sqrt{(1 - r)(1 - r)}} = \frac{1}{1 - r}$$

while at $\Omega = \pi$ since we have $\cos \pi = -1$,

$$|X(\Omega)| = \frac{1}{\sqrt{1 + 2r + r^2}} = \frac{1}{\sqrt{(1 + r)(1 + r)}} = \frac{1}{1 + r}$$

At $\Omega = \pm 2\pi$, we return to

$$|X(\Omega)| = \frac{1}{\sqrt{1 - 2r + r^2}} = \frac{1}{\sqrt{(1 - r)(1 - r)}} = \frac{1}{1 - r}$$

The Discrete Fourier Transform

In this section we consider sampling of a continuous time signal $x(t)$ that is of finite duration. We sample the signal at intervals of T_S called the *sampling period*. If we take a total of N samples of the signal, then we will have the sampled values

$$x(0), \ x(T_S), \ x(2T_S), \ \ldots, \ x[(N-1)\,T_S]$$

which define the values of a discrete time signal we now denote by $x\,[n]$. The *discrete Fourier transform* or DFT of $x\,[n]$ is denoted by $X\,[k]$ and is given by

$$X[k] = \sum_{n=0}^{N-1} x[n]\, e^{-j\frac{2\pi}{N}kn} \tag{7.21}$$

The inverse discrete Fourier transform is given by

$$x[n] = \frac{1}{N} \sum_{k=0}^{N-1} X[k]\, e^{j\frac{2\pi}{N}kn} \tag{7.22}$$

EXAMPLE 7-6
Find the discrete Fourier transform of the signal $x[n] = \{2, \ 0, \ -1, \ 3\}$, as shown in Fig. 7-4.

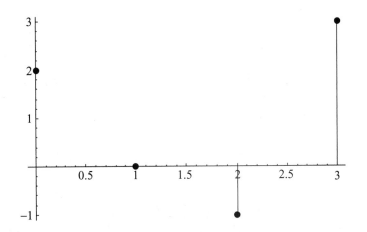

Fig. 7-4. The discrete time signal for Example 7-5.

SOLUTION 7-6

Applying (7.21) to $x[n]$, for the first term, we find

$$X[0] = \sum_{n=0}^{3} x[n] = 2 + 0 - 1 + 3 = 4$$

For the second term, we obtain

$$X[1] = \sum_{n=0}^{3} x[n] e^{-j\frac{\pi}{2}n}$$

$$= x[0] + x[1]e^{-j\frac{\pi}{2}} + x[2]e^{-j\pi} + x[3]e^{-j\frac{3\pi}{2}}$$

$$= 2 + 1 + 3j = 3 + 3j$$

Next, we have

$$X[2] = \sum_{n=0}^{3} x[n] e^{-j\pi n}$$

$$= x[0] + x[1]e^{-j\pi} + x[2]e^{-j2\pi} + x[3]e^{-j3\pi}$$

$$= 2 - 1 - 3 = -2$$

The last term in the sequence is given by

$$X[3] = \sum_{n=0}^{3} x[n] e^{-j\frac{3\pi}{2}n}$$

$$= x[0] + x[1]e^{-j\frac{3\pi}{2}} + x[2]e^{-j3\pi} + x[3]e^{-j\frac{9\pi}{2}}$$

$$= 2 + 1 - 3j = 3 - 3j$$

The real and imaginary parts of $X[k]$ are shown in Figs. 7-5 and 7-6, respectively.

EXAMPLE 7-7

Given that $X[k] = \{0, -3 - 3j, -2, -3 + 3j\}$, use the inverse discrete Fourier transform to find $x[n]$.

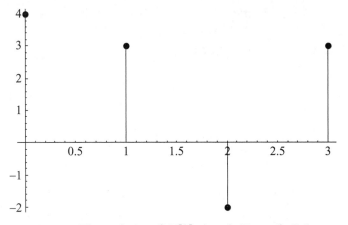

Fig. 7-5. The real part of $X[k]$ given in Example 7-5.

SOLUTION 7-7

We find the inverse discrete Fourier transform by applying (7.22). We have $N = 4$ and so

$$x[n] = \frac{1}{4} \sum_{k=0}^{3} X[k]\, e^{j\frac{\pi}{2}kn}$$

The first member of the sequence is

$$x[0] = \frac{1}{4} \sum_{k=0}^{3} X[k] = \frac{1}{4}(-3 - 3j - 2 - 3 + 3j) = \frac{1}{4}(-3 - 2 - 3) = -2$$

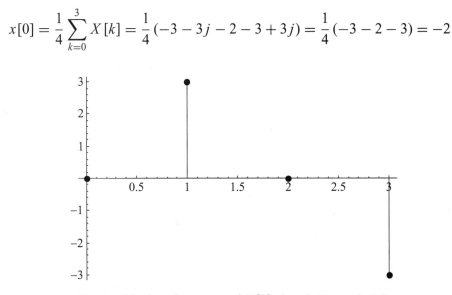

Fig. 7-6. The imaginary part of $X[k]$ given in Example 7-5.

The $n = 1$ term is given by

$$x[1] = \frac{1}{4} \sum_{k=0}^{3} X[k] \, e^{j\frac{\pi}{2}k}$$

$$= \frac{1}{4} \left[-(3+3j) \, e^{j\frac{\pi}{2}} - 2 \, e^{j\pi} + (-3+3j) \, e^{j\frac{3\pi}{2}} \right]$$

$$= \frac{1}{4} [-(3+3j) \, j - 2 \, e^{j\pi} - (-3+3j) \, j]$$

$$= \frac{1}{4} [-3j + 3 + 2 + 3j + 3]$$

$$= \frac{1}{4} (8) = 2$$

For $n = 2$, we obtain

$$x[2] = \frac{1}{4} \sum_{k=0}^{3} X[k] \, e^{j\pi k}$$

$$= \frac{1}{4} [-(3+3j) \, e^{j\pi} - 2 \, e^{j2\pi} + (-3+3j) \, e^{j3\pi}]$$

$$= \frac{1}{4} [(3+3j) - 2 - (-3+3j)]$$

$$= \frac{1}{4} [3 - 2 + 3]$$

$$= \frac{1}{4} (4) = 1$$

Finally, setting $n = 3$ yields

$$x[3] = \frac{1}{4} \sum_{k=0}^{3} X[k] \, e^{j\frac{3\pi}{2}k}$$

$$= \frac{1}{4} \left[-(3+3j) \, e^{j\frac{3\pi}{2}} - 2 \, e^{j3\pi} + (-3+3j) \, e^{j\frac{9\pi}{2}} \right]$$

$$= \frac{1}{4} [(3+3j) \, j + 2 + (-3+3j) \, j]$$

$$= \frac{1}{4} [3j - 3 + 2 - 3j - 3]$$

$$= \frac{1}{4} (-4) = -1$$

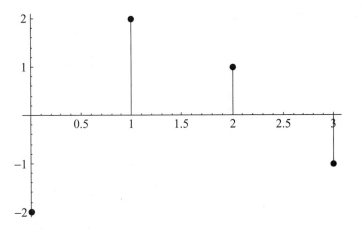

Fig. 7-7. The sequence found using the inverse discrete Fourier transform in Example 7-7.

This sequence is shown in Fig. 7-7.

   In many cases, it is helpful to once again call upon the geometric series to help in the evaluation of the discrete Fourier transform. In the case of a finite sum, the following result holds:

$$\sum_{n=0}^{N-1} ax^n = \frac{a\left(1 - x^N\right)}{1 - x} = \frac{a}{1 - x} - \frac{ax^N}{1 - x} \tag{7.23}$$

EXAMPLE 7-8
Find the discrete Fourier transform of $x[n] = a^n$ where a is a constant and $0 \le n \le N - 1$.

SOLUTION 7-8
Applying (7.21), we have

$$X[k] = \sum_{n=0}^{N-1} x[n] e^{-j\frac{2\pi}{N}kn} = \sum_{n=0}^{N-1} a^n e^{-j\frac{2\pi}{N}kn}$$

for $0 \le k \le N - 1$. Now we use $x^r y^r = (xy)^r$ to write this as

$$X[k] = \sum_{n=0}^{N-1} \left(a e^{-j\frac{2\pi}{N}k}\right)^n$$

Now we can apply (7.23) to obtain the result

$$X[k] = \frac{1 - a^N e^{-j2\pi k}}{1 - a e^{-j2\pi k/N}} = \frac{1 - a^N}{1 - a e^{-j2\pi k/N}}$$

To obtain the last step, we used
$$e^{-j2\pi k} = \cos(2\pi k) - j \sin(2\pi k) = \cos(2\pi k) = +1$$

Sampling in Detail

Now let's examine the problem of sampling a signal in more detail. For our first case, let's consider a sine wave with a frequency of 250 Hz. Using $f = 250$ Hz, we find that the period of the signal is

$$T = \frac{1}{f} = \frac{1}{250} = 4 \text{ ms}$$

The continuous time signal that we will sample is therefore defined by

$$x(t) = \sin(2\pi f t) = \sin[2\pi (250) t] \tag{7.24}$$

We show this signal over a period from $0 \le t \le 24$ ms in Fig. 7-8.

To investigate the effect of sampling at different rates, let's examine the signal over a period, which is 4 ms. This is shown in Fig. 7-9.

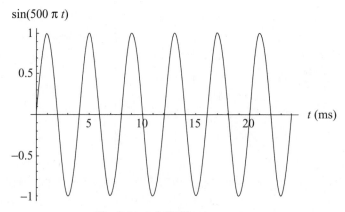

Fig. 7-8. A 250-Hz sine wave.

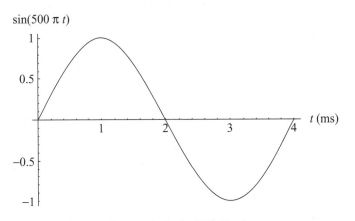

$\sin(500\,\pi\,t)$

Fig. 7-9. One period of a 250-Hz sine wave.

Now let's suppose that our sampling rate is 5000 Hz. This means that the sampling interval is

$$\Delta t = \frac{1}{5000 \text{ Hz}} = 0.2 \text{ ms}$$

Over the range from $0 \le t \le 24$ ms means that we will take $N = 120$ samples, giving us $T_{\text{S}} = N\Delta t = 24$ ms. Looking at the sampled values out to 1 ms, we have the discrete time signal

$$x[n] = \{0, 0.3090, 0.5878, 0.8090, 0.9511, 1\}$$

A plot of the discrete time signal over the entire period is shown in Fig. 7-10. As you can see, it approximates the actual signal shape fairly well.

Now let's cut the sampling rate in half, to 2500 Hz. This means the sampling interval becomes

$$\Delta t = \frac{1}{2500 \text{ Hz}} = 0.4 \text{ ms}$$

The shape of the signal begins to be more and more distorted as we decrease the sampling rate. We show the discrete time signal, in Fig. 7-11, we obtain from sampling at 2500 Hz, this time joining the plot points to get an idea of the shape of the signal.

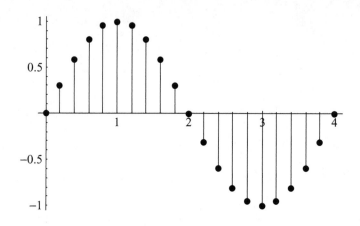

Fig. 7-10. Sampling at 5000 Hz.

Now let's cut the sampling rate to 1700 Hz. This gives us a sampling interval of about 0.59 ms. As shown in Fig. 7-12, the sampled signal shape is drifting further away from the actual signal shape.

Now let's look again at the sampling, this time cutting our sampling rate to 900 Hz. This gives a sampling interval of about 1.11 ms. As you can see in Fig. 7-13, a great deal of information about the signal has been lost.

Going one step further, in Fig. 7-14, we sample the signal at 400 Hz. This little exercise emphasizes a few basic facts:

- If a signal is changing rapidly in time, then the sampling interval Δt must be small enough to capture this fact.

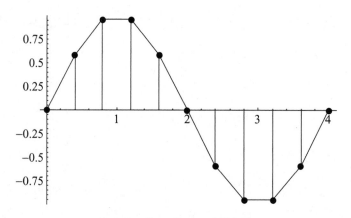

Fig. 7-11. Sampling at 2500 Hz.

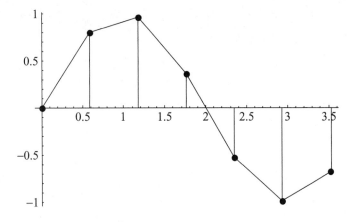

Fig. 7-12. Cutting the sampling rate to 1700 Hz causes more problems. The variation of the signal is not being captured correctly by our sampling.

- High-frequency variation can mean there are high-frequency components in the signal.
- Putting the two previous points together, we infer that a signal with high-frequency components requires a higher sampling rate.

When the sampling rate is not high enough (or conversely if the sampling interval Δt is not small enough) to correctly capture the variation in the signal, we say that *aliasing* has occurred.

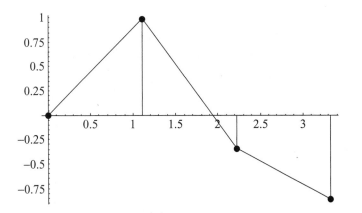

Fig. 7-13. The discrete time signal obtained from sampling a 250-Hz sine wave at 900 Hz. The signal shape has been greatly changed in this case. In this case, sampling at a small rate has caused us to lose a great deal of information about the variations in the signal.

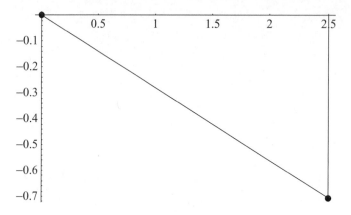

Fig. 7-14. Sampling at 400 Hz does not reveal much information about the variation in a 250-Hz sine wave.

THE NYQUIST RATE

A quantifiable way out of this dilemma is provided by the *Nyquist sampling theorem*. In words, this theorem tells us that in order to sample a signal correctly, our sampling rate should be at least twice the highest frequency component present in the signal. Let's call the highest frequency component present in a signal ω_h and our sampling rate ω_s. The Nyquist theorem tells us that to correctly resolve the signal we need to satisfy

$$\omega_s \geq 2\omega_h \tag{7.25}$$

If a signal is band limited to $[-\omega/2, \omega/2]$, then the critical sampling rate $\Delta t = 1/\omega$ is known as the *Nyquist frequency*. This tells us that if we choose our sampling rate to be $\Delta t = 1/\omega$, then ω is the highest frequency that we can pick up.

In the previous section where we sampled the sine wave at different rates, we saw that the higher our sampling rate, the better the reproduction of the signal.

LEAKAGE

Another issue that can arise when considering the sampling of signals is the problem of *leakage*. This is a phenomenon that occurs in the frequency domain, and in a nutshell what happens is that frequency components in the sampled signal $x[n]$ "leak" away from the actual frequency components found in $x(t)$.

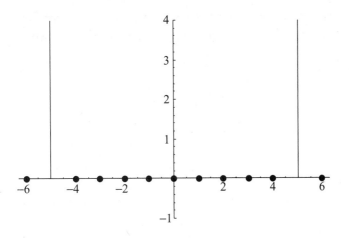

Fig. 7-15. The Fourier transform of $x(t) = \cos[2\pi (5) t]$ gives two spikes.

This problem occurs because a signal is actually sampled over a finite period of time called a *window*.

As an example of leakage, consider the Fourier transform of a cosine function. Let us suppose for the sake of example that we have $x(t) = \cos[2\pi (5) t]$. The Fourier transform of this function is two spikes located at ± 5. This is shown in Fig. 7-15.

To illustrate the phenomenon of leakage, we consider a small error in frequency. Let us suppose that instead of $x(t) = \cos[2\pi (5) t]$ we have

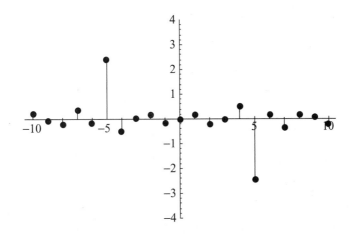

Fig. 7-16. Leakage illustrated by a small-frequency error.

$f \to 5.2$, giving us the time-domain signal $x(t) = \cos[2\pi(5.2)t]$. The discrete Fourier transform of this signal results in somewhat messy expression

$$X[k] = \frac{(-1)^k}{N}\left(\frac{\sin\left[\frac{2\pi}{N}(5.2+k)\right]}{\sin^2\left[\frac{\pi}{N}(5.2+k)\right]} - \frac{\sin\left[\frac{2\pi}{N}(-5.2+k)\right]}{\sin^2\left[\frac{\pi}{N}(-5.2+k)\right]}\right)$$

We show an example of this in Fig. 7-16. As you can see from the plot, there are several frequency components present that were not found in the signal before.

Circular Convolution

We complete our examination of the discrete Fourier transform by considering *circular convolution*. This type of convolution is the same as ordinary discrete convolution; however, it is done "mod N." The defining formula for circular convolution is given by

$$x[n] \otimes h[n] = \sum_{i=0}^{N-1} x[i]h[n-i]_{\mathrm{mod}\ N} \tag{7.26}$$

We have seen that convolution in the time domain is converted into multiplication in the frequency domain. This is also true in this case. We have the discrete Fourier transform pair

$$x[n] \otimes h[n] \rightleftharpoons X[k]H[k] \tag{7.27}$$

Therefore using the discrete Fourier transform, we can compute the circular convolution $y[n] = x[n] \otimes h[n]$ in the following steps:

- compute the discrete Fourier transforms $X[k]$ and $H[k]$;
- pointwise multiply these sequences, giving $Y[k] = X[k]H[k]$; and
- compute the inverse discrete Fourier transform of $Y[k]$ to obtain the output signal $y[n]$.

It's best to demonstrate the application of (7.26) with an example.

EXAMPLE 7-9
Find the circular convolution of

$$x[n] = \cos\left(\frac{\pi}{4}n\right) \quad \text{and} \quad h[n] = \sin\left(\frac{\pi}{2}n\right)$$

for $n = 0, 1, 2, 3$.

SOLUTION 7-9
First let's explicitly write out the values of these sequences. In the case of

$$x[n] = \cos\left(\frac{\pi}{4}n\right)$$

we have

$$x[0] = \cos(0) = 1$$

$$x[1] = \cos\left(\frac{\pi}{4}\right) = \frac{1}{\sqrt{2}}$$

$$x[2] = \cos\left(\frac{\pi}{2}\right) = 0$$

$$x[3] = \cos\left(\frac{3\pi}{4}\right) = -\frac{1}{\sqrt{2}}$$

For

$$h[n] = \sin\left(\frac{\pi}{2}n\right)$$

we have

$$h[0] = 0, \qquad h[1] = 1$$
$$h[2] = 0, \qquad h[3] = -1$$

Now we apply (7.26) to these signals. First, we flip $h[n]$ about the origin, and then shift. This gives the sequence $\{h[0], h[3], h[2], h[1]\}$.
Using $x[n] \otimes h[n] = \sum_{i=0}^{N-1} x[i]h[n-i]_{\bmod N}$, we have

$$y[0] = x[0]h[0] + x[1]h[3] + x[2]h[2] + x[3]h[1]$$
$$= (1)(0) + (1/\sqrt{2})(-1) + 0 + (-1/\sqrt{2})(1) = -\sqrt{2}$$

Now we shift h by 1 to compute the next term as

$$y[1] = x[0]h[1] + x[1]h[0] + x[2]h[3] + x[3]h[2]$$
$$= (1)(1) + (1/\sqrt{2})(0) + 0 + (-1/\sqrt{2})(0) = 1$$

The shifting procedure is applied again to compute the next term as

$$y[2] = x[0]h[2] + x[1]h[1] + x[2]h[0] + x[3]h[3]$$
$$= (1)(0) + (1/\sqrt{2})(1) + 0 + (-1/\sqrt{2})(-1) = \sqrt{2}$$

We shift once more to obtain the last term as

$$y[3] = x[0]h[3] + x[1]h[2] + x[2]h[1] + x[3]h[0]$$
$$= (1)(-1) + (1/\sqrt{2})(0) + 0 + (-1/\sqrt{2})(0) = -1$$

So we have found that

$$y[n] = \{-\sqrt{2},\ 1,\ \sqrt{2},\ -1\} \tag{7.28}$$

Now we compute the circular convolution by multiplication in the frequency domain. Using $X[k] = \sum_{n=0}^{N-1} x[n]\, e^{-j\frac{2\pi}{N}kn}$, the discrete Fourier transform of $x[n]$ is

$$X[0] = \sum_{n=0}^{3} x[n] = 1 + \frac{1}{\sqrt{2}} - \frac{1}{\sqrt{2}} = 1$$

$$X[1] = \sum_{n=0}^{3} x[n]\, e^{-j\frac{\pi}{2}n} = 1 + \frac{1}{\sqrt{2}}e^{-j\frac{\pi}{2}} - \frac{1}{\sqrt{2}}e^{-j\frac{3\pi}{2}} = 1 - j\sqrt{2}$$

$$X[2] = \sum_{n=0}^{3} x[n]\, e^{-j\pi n} = 1 + \frac{1}{\sqrt{2}}e^{-j\pi} - \frac{1}{\sqrt{2}}e^{-j3\pi} = 1$$

$$X[3] = \sum_{n=0}^{3} x[n]\, e^{-j\frac{3\pi}{2}n} = 1 + \frac{1}{\sqrt{2}}e^{-j\frac{3\pi}{2}} - \frac{1}{\sqrt{2}}e^{-j\frac{9\pi}{2}} = 1 + j\sqrt{2}$$

The discrete Fourier transform of $h[n]$ is given by

$$H[k] = \{0, \ -2j, \ 0, \ 2j\} \tag{7.29}$$

To compute the circular convolution, we apply (7.27) and pointwise multiply these sequences, thus obtaining

$$Y[0] = X[0]H[0] = (1)(0) = 0$$
$$Y[1] = X[1]H[1] = (1 - j\sqrt{2})(-2j) = -2\sqrt{2} - 2j$$
$$Y[2] = X[2]H[2] = (1)(0) = 0$$
$$Y[3] = X[3]H[3] = (1 + j\sqrt{2})(2j) = -2\sqrt{2} + 2j$$

It can be shown that computing the inverse discrete Fourier transform of this sequence yields (7.28).

Quiz

In problems 1 and 2, consider the discrete time sequence as shown in Fig. 7-17.

1. Write down the sequence $x[n]$, fundamental period N_0, and the fundamental frequency Ω_0 for the sequence shown in Fig. 7-17.
2. Find the spectral coefficients for this sequence.
3. Find the discrete Fourier transform of $x[n] = \{-2, \ 4, \ 1, \ 5\}$.

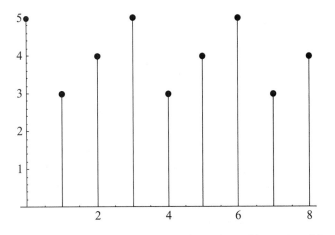

Fig. 7-17. Discrete time sequence for quiz problems 1 and 2.

4. If $X[k] = \{2, \ 2 - 1.7321j, \ 5, \ 2, \ 5, \ 2 + 1.7321j\}$, find $x[n]$.

5. Using (7.14), find the Fourier transform of a discrete time square pulse that ranges from -3 to 3, and plot it.

6. Given that

$$x[n] = \cos\left(\frac{\pi}{2}n\right)$$

and $h[n] = \{1, \ 0, \ 0, \ -1\}$ with $N = 4$, compute $y[n] = x[n] \otimes h[n]$.

CHAPTER 8

Amplitude Modulation

We now turn our attention to a central task in communications—the transmission of an information-bearing signal over a bandpass communications channel. Let us denote the information-bearing signal by $m(t)$, where m tells us that this is some type of *message signal.* Generally speaking, the transmission of such a signal requires some sort of manipulation. To see why, consider the everyday situation of AM talk radio. A radio host at the station is speaking, generating a message signal that has its natural frequencies in the audio range. This range of frequencies is not directly compatible with the frequencies used in radio transmission, so the signal must be modified in some way in order to get it in a form that can be transmitted. What happens is the frequency range is *shifted* to one that is suitable for the given communications channel. This is done using *modulation.*

Specifically, modulation is defined as a process by which some characteristic of a carrier signal $x_c(t)$ is varied in accordance with a modulating signal $m(t)$. The modulating signal is the actual information-bearing signal that we wish to transmit.

In this chapter we begin our study of modulation by examining perhaps the most familiar, if not the simplest, method—*amplitude modulation* (AM). This is a specific type of *continuous wave modulation* that can be used to manipulate a signal. A continuous wave carrier signal is a sinusoidal wave that we can write as

$$x_c(t) = A(t) \cos[\omega_c t + \phi(t)] \tag{8.1}$$

This signal includes

- $A(t)$, which is the *instantaneous amplitude*.
- $\phi(t)$, which is the *instantaneous phase angle*.
- $\omega_c = 2\pi f_c$, which is the *carrier frequency*.

Amplitude Modulation

In this chapter we study the specific case of *amplitude modulation*. This means that the instantaneous amplitude $A(t)$ of the carrier signal $x_c(t)$ is *linearly* related to the message signal $m(t)$. We take the amplitude of the carrier signal to be a constant called the *carrier amplitude*, which we denote by A_c. Furthermore, for simplicity we set $\phi(t) = 0$, which can be done without any loss of generality. This allows us to write the carrier signal as

$$x_c(t) = A_c \cos(\omega_c t) \tag{8.2}$$

Types of Amplitude Modulation

There are several types of amplitude modulation and we will consider each type in turn. These are

- Standard or *ordinary amplitude modulation*
- Double-sideband modulation (DSB)
- Single-sideband modulation (SSB)
- Vestigal-sideband modulation (VSB)

ORDINARY AMPLITUDE MODULATION

An ordinary AM signal can be created in three steps. First we define the carrier signal, which is shown in (8.2). Next, we multiply the message signal $m(t)$ by $\cos(\omega_c t)$ to give

$$m(t) \cos(\omega_c t) \tag{8.3}$$

We then form the sum of these two waves to produce the ordinary AM signal, which we denote by $x_{AM}(t)$:

$$x_{AM}(t) = m(t)\cos(\omega_c t) + A_c\cos(\omega_c t) = [A_c + m(t)]\cos(\omega_c t) \qquad (8.4)$$

The envelope

Notice that the amplitude of the ordinary AM wave $x_{AM}(t)$ is given by

$$a(t) = A_c + m(t) \quad \text{(the envelope)} \qquad (8.5)$$

This function is called the *envelope* of the AM wave. The envelope can be used to quantify the quality of the wave in regards to the transmission of the message signal in the following way. We define the *modulation index*, which is given by

$$\mu = \frac{|\max\{m(t)\}|}{A_c} \qquad (8.6)$$

This quantity indicates the amount of variation of the modulated signal about its normal value. Our concern in analyzing an AM wave will be to ensure that $\mu \leq 1$. This leads to two general cases of study.

Case (a): $\mu \leq 1$
　　　　If $\mu \leq 1$, then the envelope of $x_{AM}(t)$ will have a direct correspondence with the message signal. Such a wave can be *demodulated*, allowing the recovery of the original message signal $m(t)$.

Case (b): $\mu > 1$
　　　　If $\mu > 1$, this indicates a problem. In this case we say that the wave is *overmodulated*. Specifically, in this case the envelope of $x_{AM}(t)$ will not always directly correspond to the message signal $m(t)$. In other words the signal suffers from *envelope distortion*.

　　In some contexts, you will see the envelope written as

$$a(t) = A_c[1 + k_a m(t)]$$

where k_a is called the *amplitude sensitivity*. The same condition holds as for μ; you can see that

$$\mu = k_a |\max\{m(t)\}|$$

Therefore the above two cases correspond to $k_a|\max\{m(t)\}| \leq 1$ and $k_a|\max\{m(t)\}| > 1$ respectively. If we multiply by 100, we obtain the *percent modulation*:

$$k_a|\max\{m(t)\}|(100) \tag{8.7}$$

EXAMPLE 8-1

In this example we construct a hypothetical signal with values chosen simply for illustration. Let

$$m(t) = \sin\left(\frac{\pi}{2}t\right)$$

and describe the AM wave generated when the carrier wave is given by $x_c(t) = 3\cos(20\pi t)$.

SOLUTION 8-1

In Fig. 8-1, we show the hypothetical message signal given by $m(t) = \sin\left(\frac{\pi}{2}t\right)$.

To construct the AM signal, looking at the carrier signal, we have a carrier amplitude given by $A_c = 3$ and $f_c = 10$ Hz. Considering the angular frequency, the carrier frequency is

$$\omega_c = 2\pi f = 20\pi \text{ rad/s}$$

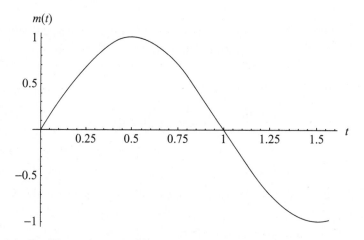

Fig. 8-1. For illustration, consider a message signal given by $m(t) = \sin\left(\frac{\pi}{2}t\right)$.

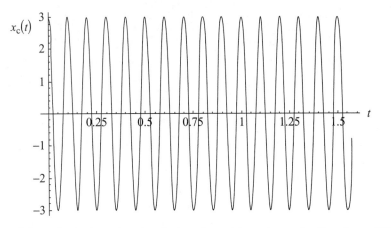

Fig. 8-2. A hypothetical carrier signal given by $x_c(t) = A_c \cos(\omega_c t) = 3 \cos(20\pi t)$.

The carrier signal is given by

$$x_c(t) = A_c \cos(\omega_c t) = 3 \cos(20\pi t)$$

This signal is shown in Fig. 8-2.

Notice that the value of the carrier signal varies as $-3 \le x_c \le 3$. Now, we construct an AM signal by adding $m(t) \cos(20\pi t) = \sin(\frac{\pi}{2}t) \cos(20\pi t)$ to the carrier signal. This gives us

$$x_{AM}(t) = \left[3 + \sin\left(\frac{\pi}{2}t\right) \right] \cos(20\pi t) \qquad (8.8)$$

The envelope is given by

$$a(t) = 3 + \sin\left(\frac{\pi}{2}t\right) \qquad (8.9)$$

Before we plot the AM signal, let's plot the envelope by itself. This is simple enough: it is just the message signal shifted upward by $A_c = 3$. This is shown in Fig. 8-3.

When plotting an AM wave, you will see that the negative values of the wave are also constrained by the envelope. To show this we can plot the negative of the envelope. In Fig. 8-4, we show a plot of $\pm a(t)$ for the present case.

Finally, in Fig. 8-5, we show that $x_{AM}(t) = [3 + \sin(\frac{\pi}{2}t)] \cos(20\pi t)$, together with the envelope. Notice how the AM wave has a one-to-one correspondence with the message signal.

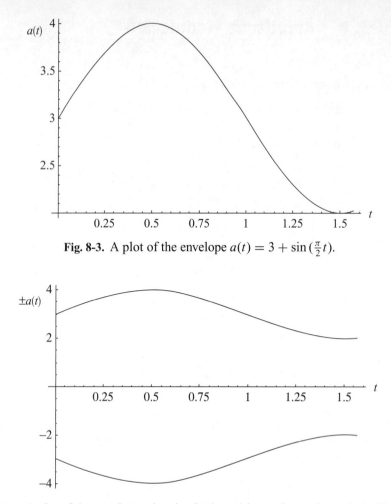

Fig. 8-3. A plot of the envelope $a(t) = 3 + \sin\left(\frac{\pi}{2}t\right)$.

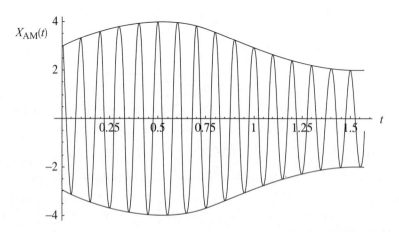

Fig. 8-4. A plot of the envelope, showing both positive and negative values. The AM wave will fit between these two curves.

Fig. 8-5. A plot of the hypothetical AM wave used in Example 8-1.

Avoiding Envelope Distortion

Two basic conditions must be met so that the AM wave $x_{AM}(t)$ does not suffer from envelope distortion. We want the envelope of $x_{AM}(t)$ to have a one-to-one correspondence with the message signal $m(t)$. For this to happen

- $\mu \leq 1$ so that the signal does not suffer from envelope distortion.
- The message bandwidth should be small compared to the carrier frequency ω_c.

We can find a nice expression for the modulation index by considering the maximum and minimum values of the envelope. Let us denote these values by a_{max} and a_{min}. Then the modulation index is given by

$$\mu = \frac{a_{max} - a_{min}}{a_{max} + a_{min}} \tag{8.10}$$

In Example 8-1, the envelope was given by

$$a(t) = 3 + \sin\left(\frac{\pi}{2}t\right)$$

Since the maximum of $\sin\left(\frac{\pi}{2}t\right)$ is $+1$ and the minimum of $\sin\left(\frac{\pi}{2}t\right)$ is -1, we have

$$a_{max} = 3 + 1 = 4$$

for the maximum and

$$a_{min} = 3 - 1 = 2$$

for the minimum. The modulation index is

$$\mu = \frac{4-2}{4+2} = \frac{2}{6} = \frac{1}{3}$$

which corresponds to a percent modulation of 33.3%.

EXAMPLE 8-2

Consider *single-tone modulation*, where the modulating signal is given by

$$m(t) = A_m \cos\left(\frac{3\pi}{4}t\right)$$

Consider the same carrier wave used in Example 8-1, that is $x_c(t) = A_c \cos(\omega_c t) = 3 \cos(20\pi t)$. Write the modulating signal in terms of the modulation index and then consider three cases of percent modulation: (a) 15%, (b) 40%, and (c) 125%.

SOLUTION 8-2

Using (8.6) we can express the amplitude of the modulating signal in terms of the modulation index as

$$A_m = A_c \mu$$

For Case (a), $\mu = 0.15$ and so $A_m = A_c \mu = (3)(0.15) = 0.45$. The AM signal in this case is

$$x_{AM}(t) = \left[3 + (0.45) \cos\left(\frac{3\pi}{4}t\right)\right] \cos(20\pi t) \tag{8.11}$$

A plot of this wave is shown in Fig. 8-6.

In Case (b), $\mu = 0.4$ and so $A_m = A_c \mu = (3)(0.4) = 1.2$. The AM signal in this case is

$$x_{AM}(t) = \left[3 + (1.2) \cos\left(\frac{3\pi}{4}t\right)\right] \cos(20\pi t) \tag{8.12}$$

This wave is shown in Fig. 8-7.

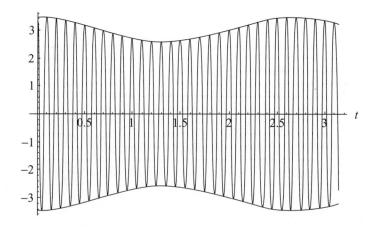

Fig. 8-6. The AM signal for the case of $\mu = 0.15$ in Example 8-2. The AM wave is that given in (8.11).

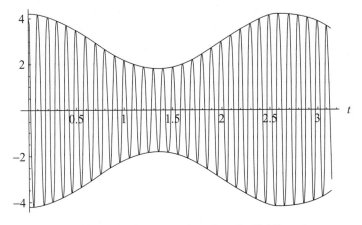

Fig. 8-7. The AM wave given in (8.12).

In Cases (a) and (b), the condition $\mu \leq 1$ was satisfied, and so there was no envelope distortion. In Case (c) however, we are considering $\mu = 1.25$. Let's examine what happens in this case.

With $\mu = 1.25$, we have $A_{\mathrm{m}} = A_{\mathrm{c}}\mu = (3)(1.25) = 3.75$. The AM wave is then

$$x_{\mathrm{AM}}(t) = \left[3 + (3.75)\cos\left(\frac{3\pi}{4}t\right) \right] \cos\left(20\pi t\right) \tag{8.13}$$

This wave is shown in Fig. 8-8. Notice that this time the shape of the message signal is not preserved for all time, and this signal suffers from *envelope distortion*.

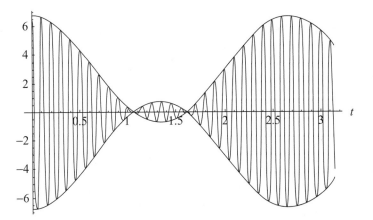

Fig. 8-8. The signal given by (8.13) provides an example of envelope distortion with $\mu = 1.25$.

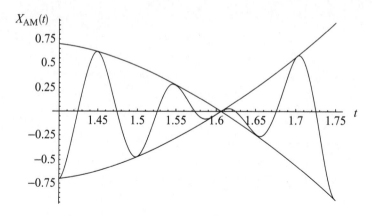

Fig. 8-9. The phase reversal for Case (c) in Example 8-2.

Let's examine this signal more closely. In Fig. 8-9, a close-up of the wave reveals that there have been two *phase reversals*, which have distorted the envelope of the wave. This is a characteristic that is seen for percent modulation that exceeds 100%.

EXAMPLE 8-3
Suppose that a message signal is given by

$$m(t) = \frac{t}{1 + t^8}$$

(a) Plot the signal.
(b) Show the envelope corresponding to this signal.
(c) Compute $x_{AM}(t)$ for 70% modulation, assuming that the carrier frequency is $f_c = 1$ Hz.

SOLUTION 8-3
(a) In Fig. 8-10 we show the message signal between 0 and 10 s.
(b) To show the envelope, we need to figure out the carrier amplitude. We are told that the signal has 70% modulation, which means that $\mu = 0.7$. So we have

$$\frac{1}{A_c} = 0.7$$

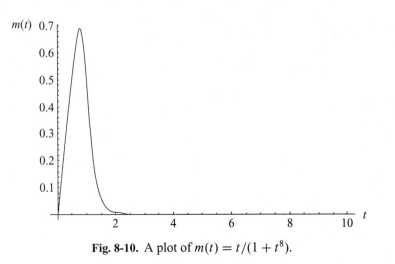

$m(t)$ 0.7

Fig. 8-10. A plot of $m(t) = t/(1 + t^8)$.

Inverting leads to $A_c = 1.43$. The envelope is then

$$a(t) = A_c + m(t) = 1.43 + \frac{t}{1 + t^8}$$

We plot the envelope showing $\pm a(t)$ in Fig. 8-11.

(c) Using what we have learned so far, the AM signal is

$$x_{AM}(t) = [A_c + m(t)]\cos(2\pi t) = \left[1.43 + \frac{t}{1 + t^8}\right]\cos(2\pi t)$$

A plot of this function is shown in Fig. 8-12.

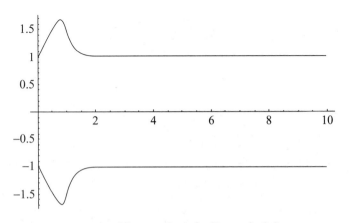

Fig. 8-11. The envelope for Example 8-3.

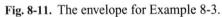

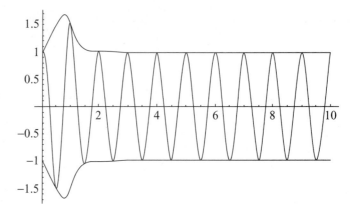

Fig. 8-12. The AM signal used in Example 8-3.

Power in AM Waves

A measure of the power contained in an AM wave is found by considering the power delivered by the wave to a single 1-Ω resistor. The *carrier power* is given in terms of the square of the carrier amplitude:

$$P_c = \frac{1}{2}A_c^2 \tag{8.14}$$

When we examine AM waves in the frequency domain, we'll learn about "sidebands." *Sideband power* is given by

$$P_s = \frac{1}{4}\mu^2 A_c^2 \tag{8.15}$$

The *total power* in the wave is

$$P_t = P_c + P_s = \frac{1}{2}\left(1 + \frac{1}{2}\mu^2\right)A_c^2 \tag{8.16}$$

The *efficiency* of an ordinary AM wave is given by the ratio of sideband power to total power in the wave. Denoted by η, the efficiency is given by

$$\eta = \frac{P_s}{P_t} \times 100\% \tag{8.17}$$

We can write the efficiency in terms of the modulation index using (8.16). This gives

$$\eta = \frac{\mu^2}{2 + \mu^2} \times 100\% \qquad (8.18)$$

EXAMPLE 8-4
Consider a System A with percent modulation equal to 50% and a System B with percent modulation equal to 90%. Find the efficiency in each case.

SOLUTION 8-4
Using (8.18), for System A with $\mu = 0.5$ we find

$$\eta_A = \frac{(0.5)^2}{2 + (0.5)^2} \times 100\% = 11.1\%$$

For System B, $\mu = 0.9$ and the efficiency is

$$\eta_B = \frac{(0.9)^2}{2 + (0.9)^2} \times 100\% = 28.8\%$$

These results mean that in System A, 11.1% of the total power is carried in the sidebands, while in System B, 28.8% of the power is carried in the sidebands.

AM Waves in the Frequency Domain

To describe the spectrum of an AM signal we first recall that the Fourier transform of a cosine function is given by

$$\cos(\omega_0 t) \rightleftharpoons \pi[\delta(\omega - \omega_0) + \delta(\omega + \omega_0)] \qquad (8.19)$$

To find the Fourier transform of an AM signal, we begin by reproducing (8.4) here, written in a convenient form:

$$x_{AM}(t) = m(t) \cos(\omega_c t) + A_c \cos(\omega_c t)$$

Referring to (8.19), the Fourier transform of the second term is immediate

$$FT[A_c \cos(\omega_c t)] = \pi A_c[\delta(\omega - \omega_c) + \delta(\omega + \omega_c)]$$

The Fourier transform of the second term can be found using the *modulation theorem*. This tells us that if the Fourier transform of $m(t)$ is given by $M(\omega)$, then the Fourier transform of $m(t) \cos \omega_c t$ is given by

$$\frac{1}{2} M(\omega - \omega_c) + \frac{1}{2} M(\omega + \omega_c) \tag{8.20}$$

To see why this is the case isn't all that difficult. Since we are working with angular frequency ω, the Fourier transform can be written as

$$X(\omega) = \int_{-\infty}^{\infty} x(t) e^{-j\omega t} \, dt \tag{8.21}$$

Let's compute the Fourier transform of $m(t) \cos \omega_c t$ directly. It will be helpful to use Euler's formula to rewrite the cosine term

$$\cos \omega_c t = \frac{e^{j\omega_c t} + e^{-j\omega_c t}}{2} \tag{8.22}$$

So the Fourier transform of the product is

$$\int_{-\infty}^{\infty} m(t) \cos \omega_c t \, e^{-j\omega t} \, dt = \int_{-\infty}^{\infty} m(t) \left(\frac{e^{j\omega_c t} + e^{-j\omega_c t}}{2} \right) e^{-j\omega t} \, dt$$

$$= \frac{1}{2} \int_{-\infty}^{\infty} m(t) e^{-j(\omega + \omega_c)t} \, dt + \frac{1}{2} \int_{-\infty}^{\infty} m(t) e^{-j(\omega - \omega_c)t} \, dt$$

The result follows. Therefore, the Fourier transform of

$$x_{AM}(t) = m(t) \cos(\omega_c t) + A_c \cos(\omega_c t)$$

is given by

$$\pi A_c [\delta(\omega - \omega_c) + \delta(\omega + \omega_c)] + \frac{1}{2} M(\omega - \omega_c) + \frac{1}{2} M(\omega + \omega_c) \tag{8.23}$$

In Fig. 8-13, we show the Fourier transform of some message signal. We say that the spectral range of the message signal is the *baseband*. Frequently the message signal is known as the baseband signal.

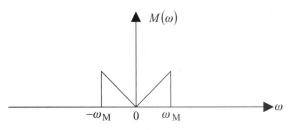

Fig. 8-13. The Fourier transform of a message signal.

In Fig. 8-14, we show the effect of modulation in the frequency domain. The effect of AM in the frequency domain can be summarized as follows:

- There are two copies of the message spectrum in the signal. One is translated to $\omega = \omega_c$. The other is translated to $\omega = -\omega_c$.
- The part of the spectrum above ω_c is called the *upper sideband*.
- The part of the spectrum below ω_c is called the *lower sideband*.
- If the bandwidth of the message signal is ω_M, then typically the carrier frequency is much larger, i.e., $\omega_c \gg \omega_M$.

Generation and Detection of Ordinary AM Waves

Now let's briefly touch on methods that can be used to generate and detect ordinary AM waves. The generation of an ordinary AM wave can be accomplished using a *square law modulator*. We won't worry about the details of how one is constructed, they can be found elsewhere. Suffice it to say for our purposes that such a device includes a nonlinear device of some kind such as a diode or transistor. Given an input signal $v_1(t)$ the transfer characteristic of such a device

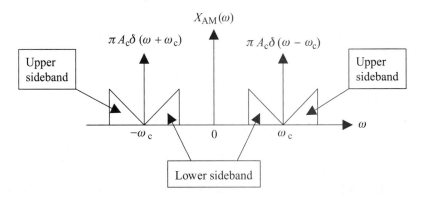

Fig. 8-14. An ordinary AM signal in the frequency domain.

is of the form

$$v_2(t) = a_1 v_1(t) + a_2 v_1^2(t) \qquad (8.24)$$

Here a_1 and a_2 are constants. The input to the device is the sum of the message signal and the carrier wave, i.e.,

$$v_1(t) = m(t) + A_c \cos \omega_c t$$

Squaring, we find

$$v_1^2(t) = [m(t) + A_c \cos \omega_c t]^2 = m^2(t) + A_c^2 \cos^2 \omega_c t + 2 A_c m(t) \cos \omega_c t$$

Therefore the output signal is given by

$$\begin{aligned}
v_2(t) &= a_1 v_1(t) + a_2 v_1^2(t) \\
&= a_1 [m(t) + A_c \cos \omega_c t] + a_2 m^2(t) + a_2 A_c^2 \cos^2 \omega_c t \\
&\quad + 2 a_2 A_c m(t) \cos \omega_c t \\
&= a_1 A_c \cos \omega_c t + 2 a_2 A_c m(t) \cos \omega_c t \\
&\quad + \{a_1 m(t) + a_2 m^2(t) + a_2 A_c^2 \cos^2 \omega_c t\}
\end{aligned}$$

The terms inside the curly braces are unwanted terms. We can get rid of them by using the appropriate filtering. The remaining term is

$$a_1 A_c \cos \omega_c t + 2 a_2 A_c m(t) \cos \omega_c t = a_1 A_c \left[1 + 2 \frac{a_2}{a_1} m(t) \right] \cos \omega_c t$$

This signal has the desired form of an AM wave.

Now let's consider a square law *detector*. The transfer characteristic is once again

$$v_2(t) = a_1 v_1(t) + a_2 v_1^2(t)$$

where a_1 and a_2 are constants. This time the input to the device is the AM wave

$$v_1(t) = x_{AM}(t) = [A_c + m(t)] \cos(\omega_c t)$$

The first term in the expression for $v_2(t)$ is just going to be the AM wave scaled by a constant. So it isn't going to be of any use in extracting the message signal.

Examining the squared term, we have

$$v_1^2(t) = x_{AM}^2(t) = [A_c + m(t)]^2 \cos^2(\omega_c t)$$

Now

$$[A_c + m(t)]^2 = A_c^2 + 2A_c m(t) + m^2(t)$$

Therefore,

$$v_1^2(t) = [A_c^2 + 2A_c m(t) + m^2(t)] \cos^2(\omega_c t)$$

$$= A_c^2 \cos^2(\omega_c t) + 2A_c \cos^2(\omega_c t)m(t) + m^2(t) \cos^2(\omega_c t)$$

So the output $v_2(t)$ assumes the form

$$v_2(t) = a_1[A_c + m(t)] \cos(\omega_c t) + a_2 A_c^2 \cos^2(\omega_c t)$$

$$+ 2a_2 A_c \cos^2(\omega_c t)m(t) + a_2 m^2(t) \cos^2(\omega_c t)$$

Looking at this expression it doesn't look like there is much hope of recovering the message signal, without a lot of further manipulation of some kind. However we focus on the term

$$2a_2 A_c \cos^2(\omega_c t)m(t)$$

Using a familiar trig identity, this can be rewritten as

$$2a_2 A_c \cos^2(\omega_c t)m(t) = 2a_2 A_c \left[\frac{1 + \cos 2\omega_c t}{2} \right] m(t)$$

$$= a_2 A_c m(t) + a_2 A_c \cos(2\omega_c t)m(t)$$

Notice the presence of the term $a_2 A_c m(t)$, which is the message signal scaled by some constants. Appropriate filtering can remove the other unwanted terms, leaving us with the message signal—the desired result. Recalling that $\omega_c \gg \omega_M$, this can be done by the application of a low pass filter. Let's think about this in some detail. Looking at the output $v_2(t)$, we see that a low pass filter that excludes ω_c will remove the first term

$$a_1[A_c + m(t)] \cos(\omega_c t)$$

The remaining terms are

$$a_2 A_c^2 \cos^2(\omega_c t) + 2a_2 A_c \cos^2(\omega_c t)m(t) + a_2 m^2(t) \cos^2(\omega_c t)$$

Applying the filter to the first term will leave a leftover constant, which isn't too much of a problem:

$$a_2 A_c^2 \cos^2(\omega_c t) = a_2 A_c^2 \left[\frac{1 + \cos 2\omega_c t}{2} \right] \overset{\text{filter } \omega_c}{\longrightarrow} \frac{a_2 A_c^2}{2}$$

We have already considered the middle term. When we filter this term we obtain the message signal

$$a_2 A_c m(t) + a_2 A_c \cos(2\omega_c t)m(t) \overset{\text{filter } \omega_c}{\longrightarrow} a_2 A_c m(t)$$

Now consider the final term. First we apply the same trig identity to write

$$a_2 m^2(t) \cos^2(\omega_c t) = \frac{a_2 m^2(t)}{2} + \frac{a_2 m^2(t) \cos 2\omega_c t}{2}$$

Passing this through a low pass filter leaves us with

$$\frac{a_2 m^2(t)}{2} + \frac{a_2 m^2(t) \cos 2\omega_c t}{2} \overset{\text{filter } \omega_c}{\longrightarrow} \frac{a_2 m^2(t)}{2}$$

This is an unwanted term—we are left with the *square* of the message signal. This unwanted term is a source of distortion.

Another type of detector that can be used with an AM wave is an *envelope detector*. This type of detector can be made with a diode, resistor, and capacitor. This is shown in Fig. 8-15.

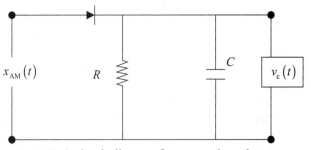

Fig. 8-15. A circuit diagram for an envelope detector.

An envelope detector works like this. In the positive half-cycle of the input signal, the diode is forward biased charging up the capacitor. When the input signal drops below the maximum, the diode shuts off and the capacitor slowly discharges into the resistor. This process is repeated as the input signal once again returns to the positive half-cycle.

In order for the simple method of envelope detection to work, it is necessary that enough carrier power be transmitted. Mathematically this can be expressed with the simple condition

$$A_c + m(t) > 0 \qquad (8.25)$$

This relation must hold for *all* values of t.

EXAMPLE 8-5
In Example 8-2 we considered the case where $A_m = A_c\mu = (3)(1.25) = 3.75$ and $A_c = 3$. Explain why this causes problems for envelope detection.

SOLUTION 8-5
Since the message signal in the problem was a cosine function, we see that at the minimum, $m(t) = -3.75$. We denote the time of a minimum by t_m. Then we have

$$A_c + m(t_m) = 3 - 3.75 = -0.75 < 0$$

Since (8.25) is not satisfied, an envelope detector would not work.

DOUBLE-SIDEBAND MODULATION

When double-sideband modulation is used, the amplitude of the AM signal is proportional to the message signal. We can write this as

$$A(t) = am(t) \qquad (8.26)$$

where a is a constant of proportionality. For simplicity, we can take $a = 1$. Then a double-sideband wave is given by

$$x_{DSB}(t) = m(t) \cos \omega_c t \qquad (8.27)$$

The generation of a DSB signal is conceptually straightforward. All that needs to be done is to multiply the message signal by the carrier wave. This is shown schematically in Fig. 8-16.

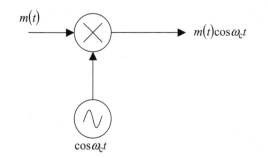

Fig. 8-16. A representation of the generation of a DSB wave.

The frequency-domain representation of this wave is easy to find using the modulation theorem. We have already been through this procedure for an ordinary AM wave, which is more complicated. The result is the same as (8.23) except the Dirac delta functions are not included since there is no stand-alone cosine term in the time-domain representation of the signal. Therefore the frequency-domain representation is given by

$$X_{\text{DSB}}(\omega) = \frac{1}{2}M(\omega - \omega_c) + \frac{1}{2}M(\omega + \omega_c) \qquad (8.28)$$

The absence of the cosine term that suppresses the appearance of the Dirac delta functions at the carrier frequency leads us to say that this is *suppressed carrier modulation.* Everything we learned about the frequency domain for an ordinary AM signal applies here other than that detail. In Fig. 8-17, we show a spectrum for a hypothetical DSB wave for the message signal in Fig. 8-13. Compare Fig. 8-17 with Fig. 8-14, which shows the spectrum for an ordinary AM wave.

Demodulation for a DSB wave is conceptually very simple. We do the same operation we did to create the signal in the first place, and then apply a low pass

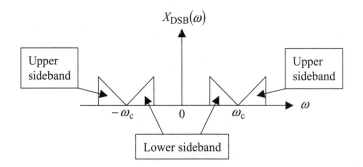

Fig. 8-17. The frequency spectrum for a DSB wave. Notice the carrier wave does not show up explicitly in the spectrum, other than via the shift in frequency of the message signal.

filter. That is we multiply $x_{DSB}(t)$ by $\cos \omega_c t$. The result is

$$x_{DSB}(t) \cos \omega_c t = m(t) \cos^2 \omega_c t = m(t) \left[\frac{1 + \cos 2\omega_c t}{2} \right]$$

Now recall that the bandwidth of the message signal, ω_M, should be much smaller than the carrier frequency, i.e., $\omega_M \ll \omega_c$. So we can apply a low pass filter that rejects the carrier frequency, and we are left with

$$m(t) \left[\frac{1 + \cos 2\omega_c t}{2} \right] \xrightarrow{\text{filter } \omega_c} \frac{1}{2} m(t)$$

This type of demodulation is called *coherent detection* because the receiver must generate a local wave that has the same frequency ω_c and is in phase with the DSB signal. As you might imagine this requirement that the receiver generate an exact wave to match the carrier leads to difficulties.

EXAMPLE 8-6
Consider a DSB receiver with a small frequency error in the local carrier. We denote the error by φ, so the local carrier wave is given by

$$\cos \left[(\omega_c + \varphi)t \right]$$

Determine the result of demodulation in this case.

SOLUTION 8-6
We assume a DSB signal of the form

$$x_{DSB}(t) = m(t) \cos (\omega_c t)$$

Demodulation with the local carrier gives the following output signal:

$$y(t) = x_{DSB}(t) \cos \left[(\omega_c + \varphi)t \right] = m(t) \cos (\omega_c t) \cos \left[(\omega_c + \varphi)t \right]$$

We recall a famous trig identity

$$\cos (x + y) = \cos x \, \cos y - \sin x \, \sin y$$

Using this expansion in $y(t)$ gives

$$y(t) = m(t) \cos (\omega_c t)[\cos (\omega_c t) \cos (\varphi t) - \sin (\omega_c t) \sin (\varphi t)]$$
$$= m(t) \cos (\varphi t) \cos^2(\omega_c t) - m(t) \cos (\omega_c t) \sin (\omega_c t) \sin (\varphi t)$$

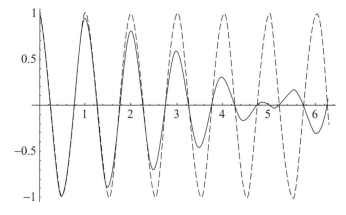

Fig. 8-18. The effect of beating due to a demodulation error.

Isolating the first term, we have

$$m(t) \cos(\varphi t) \cos^2(\omega_c t) = m(t) \cos(\varphi t) \left[\frac{1 + \cos 2\omega_c t}{2}\right]$$

$$= \frac{m(t) \cos(\varphi t)}{2} + \frac{m(t) \cos(\varphi t) \cos 2\omega_c t}{2}$$

If we pass this through a filter that rejects ω_c, then the surviving term is

$$\frac{m(t) \cos(\varphi t)}{2}$$

The presence of $\cos(\varphi t)$ indicates that there will be some distortion. The type of distortion represented by this term is known as *beating* and this is an unwanted effect. An exaggerated example of this is shown in Fig. 8-18. We examine a case of single tone modulation and assume an error term $\varphi \approx 0.05$. The original message signal is shown with the dashed line, and the signal $m(t) \cos(\varphi t)$ is shown with the solid line. Notice that as time goes on, the demodulated signal becomes highly distorted.

SINGLE-SIDEBAND MODULATION

Looking at the spectra shown in Figs. 8-14 and 8-17, you will notice that ordinary AM and DSB modulation produce two copies of the message signal in the frequency domain. This is wasted bandwidth. Since each single sideband contains complete information about the signal, it is necessary only to transmit

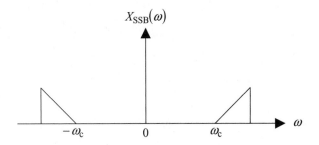

Fig. 8-19. Single-sideband modulation transmitting the upper sideband.

one of them. This type of modulation is called *single-sideband modulation*. Typically, either the lower or upper sideband is transmitted. If the upper sideband is transmitted, the resulting spectrum is that shown in Fig. 8-19.

It is also possible to generate a single-sideband signal by transmitting the lower sideband. This is shown in Fig. 8-20.

A single-sideband-modulated signal can be generated using a technique called *frequency discrimination*. Simply put, a DSB signal is created and then passed through a bandpass filter that eliminates the unwanted frequency components. This is a simple process to imagine with an ideal bandpass filter that will cut the frequency off perfectly at ω_c. But in practice this is difficult to achieve because real filters do not have the sharp cutoff that is required.

VESTIGAL-SIDEBAND MODULATION

The final modulation technique we will discuss is *vestigal-sideband modulation*. This type of modulation, which is used to transmit the television video signal, passes one sideband completely but only passes a trace of the other sideband. This is the origin of the name of this type of modulation, as the trace part of the other sideband is known as a "vestige."

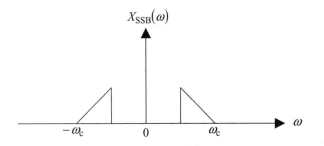

Fig. 8-20. Single-sideband modulation using the lower sideband.

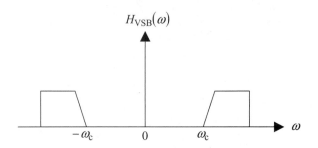

Fig. 8-21. A sample transfer function for generating VSB-modulated signals. This transfer function completely passes the upper sideband but only passes a trace of the lower sideband.

As a simple example to illustrate the concept, suppose that the transfer function for the system used to generate the VSB signal is that shown in Fig. 8-21.

This type of modulation does not save as much bandwidth as single-sideband modulation, but minimizes the sharp cutoff problem described in the last section. As an example of the frequency spectrum for the modulated signal, we once again consider the message signal shown in Fig. 8-13. This time the modulated signal has a spectrum shown in Fig. 8-22.

Frequency Translation

There are times when it is necessary to shift a signal to a different frequency band for processing. The processing could involve amplification or filtering, for example. For commercial AM, radio signals are shifted to 455 kHz, which is known as the IF or *intermediate frequency band*. The device that performs this task is called a *frequency mixer*. Suppose that the carrier frequency is ω_1. The

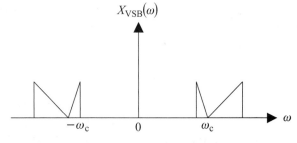

Fig. 8-22. The VSB signal.

frequency mixer produces a signal given by

$$2m(t)\cos(\omega_1 t)\cos[(\omega_1 + \omega_2)t]$$

where ω_2 is the desired local frequency. Some manipulation of this expression shows that it can be written as

$$2m(t)\cos(\omega_2 t)\left[\frac{1 + \cos 2\omega_1 t}{2}\right] - m(t)\sin 2\omega_1 t\ \sin \omega_2 t$$

This signal is passed through a filter to block ω_1, so the end result is

$$y(t) = m(t)\cos(\omega_2 t)$$

The generation of this type of signal is called *heterodyning*.

EXAMPLE 8-7
A commercial broadcast station transmits at 650 kHz. If the IF band is 455 kHz, show that the generation of the local signal results in an image frequency.

SOLUTION 8-7
The sum of the broadcast frequency and the IF frequency is $(650 + 455)$ kHz $=$ 1105 kHz. Commercial AM radio, however, ranges over 540–1600 kHz. Notice that $1105 + 455 = 1560$ kHz, which falls within this range. This is the image frequency and after heterodyning it is not possible to distinguish it from the actual signal.

Quiz

1.   If $x_c(t) = 3\cos(20\pi t)$, what is the carrier power?
2.   Suppose that a message signal is given by $m(t) = 4\cos(\pi t)$ and the carrier wave is $x_c(t) = 2\cos(40\pi t)$. Can envelope detection be used to demodulate the signal?
3.   Find the fraction of total power contained in the sidebands if there is 100% modulation.
4.   Suppose that the modulation index is $\mu = 0.08$. What fraction of power is in the sidebands?

5. In Example 8-6, the second term in the demodulated signal was ignored. Why is that? The term is $m(t) \cos(\omega_c t) \sin(\omega_c t) \sin(\varphi t)$.

6. Consider the demodulation of a DSB signal $x(t) = m(t) \cos \omega_c t$. Suppose that the local carrier wave has a phase error, such that the output is given by $y(t) = m(t) \cos(\omega_c t) \cos(\omega_c t + \pi/2)$. What is the effect of the phase error on the output?

CHAPTER 9

Angle Modulation

We are considering carrier waves that are sinusoidal. The most general carrier wave is defined by an amplitude A_c, carrier frequency ω_c, and phase angle ϕ; that is, we can write

$$x_c(t) = A_c \cos(\omega_c t + \phi) = A_c \cos(\theta)$$

This tells us that there are two basic ways to vary the carrying signal with the modulating signal. In the previous chapter, we considered amplitude modulation, in which the carrier amplitude A_c varied with the modulating signal in some way. In that case the argument of the cosine function or the *angle* θ had no relationship with the modulating signal.

The second way we could vary the carrier with the modulating signal is to let the amplitude stay constant and vary the angle with the modulating signal in some way. First notice that in this chapter we will write the angle in a more general form that includes a phase ϕ. The reader will no doubt recall that a phase shift will shift a sinusoidal signal forward or backward in time, as shown in Fig. 9-1.

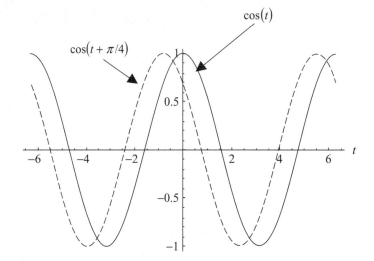

Fig. 9-1. The effect of a *phase shift*, comparing cos (t) (solid line) with cos $(t + \pi/4)$ (dashed line).

Instantaneous Frequency

We now allow the phase to vary with time, i.e., $\phi = \phi(t)$ and write the angle as

$$\theta(t) = \omega_c t + \phi(t) \tag{9.1}$$

Taking the derivative of this expression gives the *instantaneous frequency*

$$\omega_i = \frac{\infty \, \mathrm{d}\theta}{\mathrm{d}t} = \omega_c + \frac{\mathrm{d}\phi}{\mathrm{d}t} \tag{9.2}$$

The way that the instantaneous frequency is varied with the modulating signal $m(t)$ defines the type of modulation that is being used. There are two types of modulation we will consider, *phase modulation* and *frequency modulation*. In the following definitions, we will assume that the modulating signal $m(t)$ is a voltage signal. In other words, the units are

$$[m(t)] = [\text{volts}]$$

(Note: We use [ ] to denote *units of.*)

Phase Modulation Defined

In phase modulation the angle $\theta(t)$ is varied linearly with the modulating signal as follows:

$$\theta(t) = \omega_c t + k_p m(t) \qquad (9.3)$$

The constant k_p is called the *phase deviation constant*. The argument to the cosine function $\theta(t)$ has units of radians (rad); therefore, since $[m(t)] = [\text{volts}]$, it must be true that

$$[k_p] = [\text{radians/volt}]$$

Therefore, the defining equation of phase modulation is found by varying the phase of the carrier with the modulating signal

$$\phi(t) = k_p\, m(t) \qquad (9.4)$$

We will have more to say about phase modulation as we go along.

Frequency Modulation Defined

Frequency modulation is a method that varies the instantaneous frequency with the modulating signal. In this case,

$$\omega_i = \omega_c + \frac{d\phi}{dt} = \omega_c + k_f m(t) \qquad (9.5)$$

where k_f is the *frequency deviation constant*. To find the units, notice that since $[\phi] = [\text{radians}]$, $\omega_i = \omega_c + \frac{d\phi}{dt} = \omega_c + k_f m(t)[d\phi/dt] = [\text{radians/second}]$. This tells us that the units of k_f must be given by

$$[k_f] = \left[\frac{\text{radians}}{\text{volt} \cdot \text{second}}\right]$$

The defining equation of frequency modulation can be said to be

$$\frac{d\phi}{dt} = k_f\, m(t) \qquad (9.6)$$

Therefore, in the case of frequency modulation, the phase angle can be obtained by integrating (9.6) as

$$\phi(t) = k_f \int_0^t m(\tau) \, d\tau + \phi(0) \tag{9.7}$$

While we have chosen the initial time $t_0 = 0$, in many cases of interest we take $t_0 = -\infty$.

The Carrier Signal

We can now write down the carrier signal for the two cases. Using what we learned in the last two sections, the carrier signal in the case of phase modulation can be written as

$$x_{PM}(t) = A_c \cos [\omega_c t + k_p m(t)] \tag{9.8}$$

In the case of phase modulation, the angle is

$$\theta_{PM}(t) = \omega_c t + k_p m(t) \tag{9.9}$$

Therefore the instantaneous frequency of a phase-modulated wave is related to the *derivative* of the modulating signal

$$\omega_i = \omega_c + k_p \frac{dm}{dt} \quad \text{(phase-modulated wave)}$$

The carrier signal in the case of frequency modulation is given by

$$x_{FM}(t) = A_c \cos [\omega_c t + k_f \int_0^t m(\tau) \, d\tau] \tag{9.10}$$

In the case of frequency modulation, the angle is

$$\theta_{FM}(t) = \omega_c t + k_f \int_0^t m(\tau) \, d\tau \tag{9.11}$$

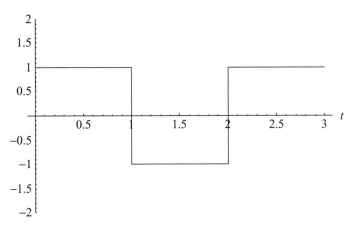

Fig. 9-2. A square wave.

Taking the derivative we find that, for frequency modulation, the instantaneous frequency is linearly related to the modulating signal as

$$\omega_i = \omega_c + k_f m(t) \quad \text{(frequency-modulated wave)}$$

One modulating signal that immediately brings out differences between phase and frequency modulation is a square wave (see Fig. 9-2).

For the sake of illustration, we set $k_p = k_f = 1$ and suppose that $\omega_c = 2\pi$ rad/s. The phase-modulated wave, plotted using (9.8), is shown in Fig. 9-3. Notice the kinks at points of phase reversal.

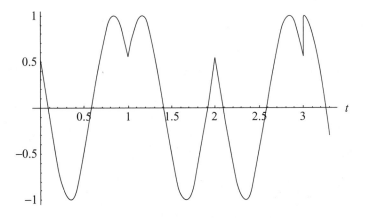

Fig. 9-3. Phase modulation for the square wave.

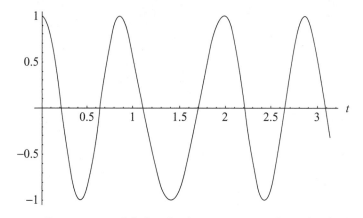

Fig. 9-4. Frequency modulation for the square wave shown in Fig. 9-2.

In Fig. 9-4, we show frequency modulation for this square wave. This wave was plotted using (9.10).

EXAMPLE 9-1
For each of the following signals, plot the function for $0 \le t \le 2\pi$, and find and plot the instantaneous frequency.

   (a) $x(t) = 2\cos(3\pi t + \pi/4)$
   (b) $x(t) = 2\cos(3\pi t + \sin(2\pi t))$
   (c) $x(t) = 2\cos(3\pi t + (\pi/4)t^3)$
   (d) $x(t) = \cos[3\pi t + 4\sin(\pi t)]\sin[3\pi t + 4\sin(\pi t)]$

SOLUTION 9-1
   (a) We show a plot of $x(t) = 2\cos(3\pi t + \pi/\pi 4)$ in Fig. 9-5. In this case,

$$\theta(t) = 3\pi t + \frac{\pi}{4}$$

We take the derivative to obtain the instantaneous frequency as

$$\frac{d\theta}{dt} = 3\pi = 2\pi\left(\frac{3}{2}\right)$$

The instantaneous frequency is 1.5 Hz and is shown in Fig. 9-6.

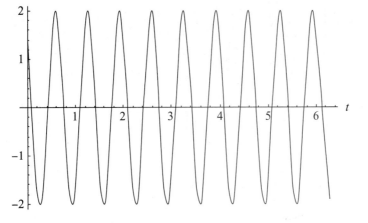

Fig. 9-5. A plot of $x(t) = 2 \cos (3\pi t + \pi/4)$.

(b) Next we have $x(t) = 2 \cos [3\pi t + \sin (2\pi t)]$. The angle is

$$\theta(t) = 3\pi t + \sin (2\pi t)$$

Taking the derivative, we obtain the instantaneous frequency as

$$\omega_i = \frac{d\theta}{dt} = 3\pi + 2\pi \cos (2\pi t) = 2\pi \left(\frac{3}{2} + \cos 2\pi t \right)$$

The instantaneous frequency is $3/2 + \cos 2\pi t$ Hz, which in this case oscillates between 0.5 and 2.5 Hz. (The numbers chosen for this exercise

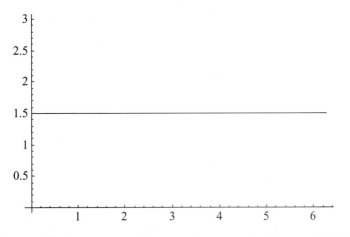

Fig. 9-6. The constant instantaneous frequency for Example 9-1(a).

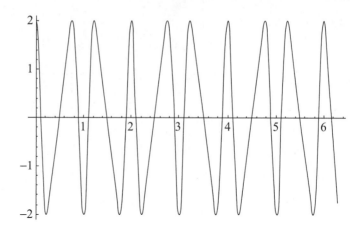

Fig. 9-7. A plot of $x(t) = 2\cos[3\pi t + \sin(2\pi t)]$.

are for pedagogical convenience.) The signal is shown in Fig. 9-7 and the instantaneous frequency in Fig. 9-8.

(c) Given that

$$x(t) = 2\cos\left(3\pi t + \frac{\pi}{4}t^3\right)$$

we have

$$\theta(t) = 3\pi t + \frac{\pi}{4}t^3$$

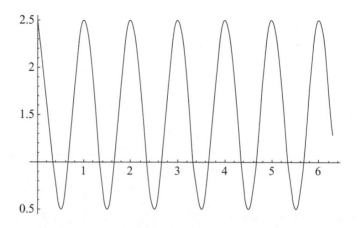

Fig. 9-8. The instantaneous frequency for $x(t) = 2\cos[3\pi t + \sin(2\pi t)]$.

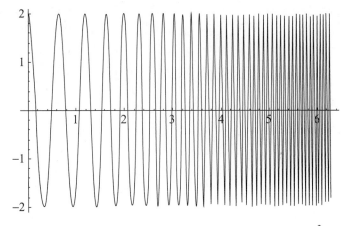

Fig. 9-9. A plot of the signal $x(t) = 2\cos[3\pi t + (\pi/4)t^3]$.

and so

$$\frac{d\theta}{dt} = 3\pi + \frac{3\pi}{4}t^2 = 2\pi\left(\frac{3}{2} + \frac{3}{8}t^2\right)$$

The signal is shown in Fig. 9-9 and the instantaneous frequency that increases with the square of t is shown in Fig. 9-10.

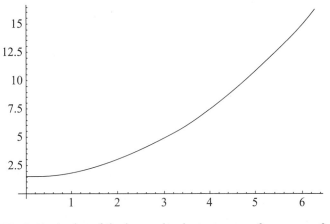

Fig. 9-10. A plot of the increasing instantaneous frequency of
$x(t) = 2\cos[3\pi t + (\pi/4)t^3]$.

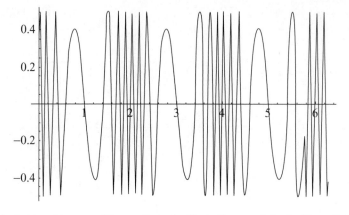

Fig. 9-11. $x(t) = \cos{[3\pi t + 4\sin{(\pi t)}]}\sin{[3\pi t + 4\sin{(\pi t)}]} = (1/2)$
$\sin{[6\pi t + 8\sin{(\pi t)}]}$.

(d) $x(t) = \cos{[3\pi t + 4\sin{(\pi t)}]}\sin{[3\pi t + 4\sin{(\pi t)}]}$

$$= \frac{1}{2}\sin{[6\pi t + 8\sin{(\pi t)}]}$$

$\theta(t) = 6\pi t + 8\sin{(\pi t)}$

$\Rightarrow \quad \dfrac{d\theta}{dt} = 6\pi + 8\pi\cos{\pi t} = 2\pi(3 + 4\cos{\pi t})$

So, the instantaneous frequency is $3 + 4\cos{\pi t}$ Hz. The signal is shown in Fig. 9-11, while in Fig. 9-12 we show the instantaneous frequency.

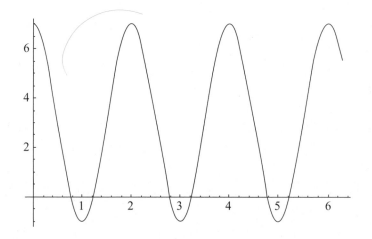

Fig. 9-12. The sinusoidally varying instantaneous frequency of Example 9-1(d).

Tone Modulation

We now consider frequency modulation where the message signal is given by

$$m(t) = A_m \cos \omega_m t \qquad (9.12)$$

Recalling (9.7), the phase of a frequency-modulated (FM) signal is found by integrating this function. Without loss of generality, we integrate from $t_0 = 0$ and obtain

$$\phi(t) = k_f \int_0^t A_m \cos \omega_m \tau \, d\tau = \frac{k_f A_m}{\omega_m} \sin \omega_m t$$

The combination $\beta = k_f A_m / \omega_m$ is called the *modulation index*. The modulation index is defined only for angle modulation when the message signal is sinusoidal. If we define the *maximum frequency deviation* as

$$\Delta \omega = \max |\omega_i - \omega_c| \qquad (9.13)$$

we have

$$\omega_i = \omega_c + \frac{d\phi}{dt} \;\Rightarrow\; \omega_i - \omega_c = \frac{d\phi}{dt} = k_f A_m \cos \omega_m t$$

And so the maximum frequency deviation is given by

$$\Delta \omega = \max |k_f A_m \cos \omega_m t| = k_f A_m$$

which allows us to write the modulation index as

$$\beta = \frac{\Delta \omega}{\omega_m}$$

The Spectrum of an FM Signal

We now consider the spectrum of an FM signal for the special case of tone modulation. First let's recall that

$$x_c(t) = A_c \cos(\omega_c t + \beta \sin \omega_m t) \qquad (9.14)$$

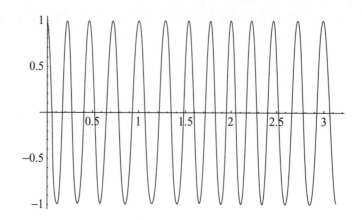

Fig. 9-13. An example of a waveform given by $x_c(t) = A_c \cos(\omega_c t + \beta \sin \omega_m t)$.

An example of this type of signal is shown in Fig. 9-13. We can rewrite this signal by recalling that

$$\cos(x + y) = \cos x \cos y - \sin x \sin y$$

Equation (9.14) then becomes

$$x_c(t) = A_c \cos(\omega_c t) \cos(\beta \sin \omega_m t) - A_c \sin(\omega_c t) \sin(\beta \sin \omega_m t)$$

This representation gives us the in-phase and *quadrature* components of the wave. The in-phase component is

$$x_I(t) = A_c \cos(\beta \sin \omega_m t)$$

The quadrature component, which is 90° out of phase, is

$$x_Q(t) = A_c \sin(\beta \sin \omega_m t)$$

If we write $\tilde{x}(t) = x_I(t) + jx_Q(t) = A_c \exp(j\beta \sin \omega_m t)$ then we can write the FM signal as

$$x_c(t) = Re\{\tilde{x}(t) e^{j\omega_c t}\} = Re\{e^{j\beta \sin \omega_m t} e^{j\omega_c t}\} = Re\{e^{j(\beta \sin \omega_m t + \omega_c)t}\} \qquad (9.15)$$

We can find a Fourier series representation of $e^{j\beta \sin \omega_m t}$. This series will be of the form

$$\tilde{x}(t) = \sum_{n=-\infty}^{\infty} c_n\, e^{jn\omega_m t}$$

The coefficients of the expansion are found from

$$c_n = \frac{1}{T_0} \int_{-T_0/2}^{T_0/2} \tilde{x}(t)\, e^{-jn\omega_m t}\, dt = \frac{\omega_m}{2\pi} \int_{-\pi/\omega_m}^{\pi/\omega_m} e^{j\beta \sin \omega_m t}\, e^{-jn\omega_m t}\, dt$$

Now we let $\omega_m t = x$ and this integral becomes

$$c_n = \frac{1}{2\pi} \int_{-\pi}^{\pi} e^{j\beta \sin x}\, e^{-jnx}\, dx = J_n(\beta)$$

where $j_n(\beta)$ is the nth-order *Bessel function of the first kind*. In Fig. 9-14, we see a plot of $J_0(\beta)$. Notice that the function oscillates between positive and negative values, and dies off as β gets large.

A few examples of higher order Bessel functions are shown in Fig. 9-15. The Bessel functions have a series representation given by

$$J_n(x) = \frac{x^n}{2^n \Gamma(n+1)} \left\{ 1 - \frac{x^2}{2(2n+2)} + \frac{x^4}{8(2n+2)(2n+4)} + \cdots \right\}$$

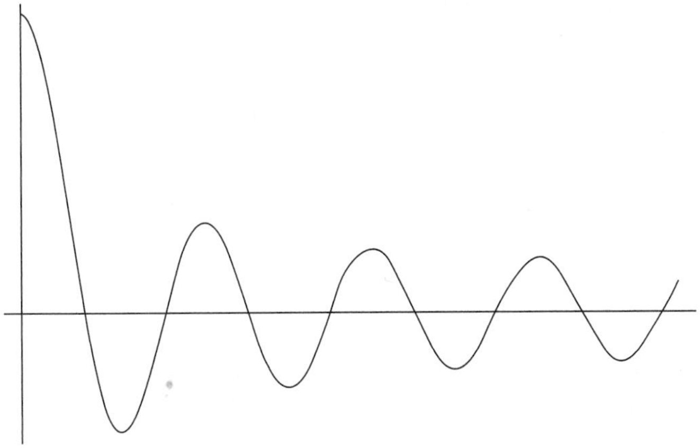

Fig. 9-14. A plot of $J_0(\beta)$.

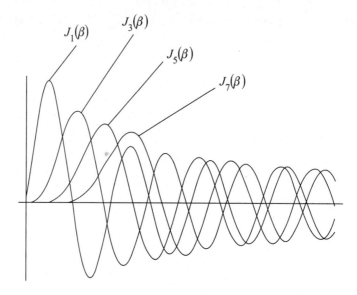

Fig. 9-15. Plot of $J_1(\beta)$, $J_3(\beta)$, $J_5(\beta)$, and $J_7(\beta)$.

Using the series representation of $J_n(\beta)$ for small β (less than a radian), we can take $J_0(\beta) \approx 1$ and $J_1(\beta) \approx \beta/2$, and all other components will be zero. This situation defines *narrowband frequency modulation*. The Bessel functions also satisfy $J_{-n}(x) = -J_n(x)$. Looking at the Fourier series representation under this condition, we have

$$\tilde{x}(t) = \sum_{n=-\infty}^{\infty} J_n(\beta)\, e^{jn\omega_m t} \approx 1 + \frac{\beta}{2} e^{j\omega_m t} - \frac{\beta}{2} e^{-j\omega_m t}$$

Using this together with (9.15) we can write the FM wave as the sum of the carrier and an upper and lower side frequency component

$$x_{FM}(t) \approx A_c \cos \omega_c t + \frac{\beta A_c}{2} \cos\left[(\omega_c + \omega_m)t\right] - \frac{\beta A_c}{2} \cos\left[(\omega_c - \omega_m)t\right] \quad (9.16)$$

The power contained in a narrowband tone-modulated signal is localized such that 98% of the total power is contained within $W_B = 2(\beta + 1)\omega_m$. In the case of a narrowband signal, where we can take $\beta \ll 1$, $W_B = 2\omega_m$.

General Series Expansion of an Angle-Modulated Signal

We have been considering tonal modulation. Let's turn to a more general situation. (Note: This includes phase as well as frequency modulation.) In general we can write $x_c(t) = \mathrm{Re}\{A_c\, e^{j\omega_c t}\, e^{j\phi(t)}\}$. The exponential term $e^{j\phi(t)}$ can be expanded in a power series, allowing us to write the signal as

$$x_c(t) = A_c\left[\cos\omega_c t - \phi(t)\sin\omega_c t - \frac{\phi^2(t)}{2!}\cos\omega_c t + \frac{\phi^3(t)}{3!}\sin\omega_c t + \cdots\right]$$

(9.17)

In this general case, the condition for narrowband frequency modulation can be expressed as

$$\max|\phi(t)| \ll 1$$

Neglecting the higher order terms in (9.17), we can write a narrowband signal as

$$x_c(t) = A_c(\cos\omega_c t - \phi(t)\sin\omega_c t)$$

(9.18)

The multiplication of the phase $\phi(t)$ by $\sin\omega_c t$ (which is 90° out of phase with the carrier) produces the two sidebands. In the case of frequency modulation for a general message signal $m(t)$, a narrowband signal can be written as

$$x_{\mathrm{NM\,FM}}(t) = A_c\cos\omega_c t - A_c\left[k_f\int_{-\infty}^{t} m(\tau)\,d\tau\right]\sin\omega_c t$$

(9.19)

In the case of phase modulation, a narrowband signal is given by

$$x_{\mathrm{NM\,PM}}(t) = A_c\cos\omega_c t - A_c k_p m(t)\sin\omega_c t$$

(9.20)

If the bandwidth of $\phi(t)$ is W_B then the bandwidth of the narrowband signal is $2W_B$.

The Deviation Ratio

We now seek a quantity that can play the role of the modulation index in a more general situation that extends beyond the case of tonal modulation. The quantity we seek is given by the *deviation ratio*, which is defined using

$$D = \frac{\Delta\omega}{\omega_m} \tag{9.21}$$

where $\Delta\omega$ is given by (9.13).

Carson's Rule

The majority of the signal power is concentrated in a bandwidth given by

$$W_B = 2(D + 1)\omega_m \tag{9.22}$$

For small D, the bandwidth is given by $W_B = 2\omega_m$.

Wideband Signals

If $\beta \gg 1$ in the case of tonal modulation or $D \gg 1$ in the more general case, we say that we have a *wideband signal*. In that case the FM signal is made up of the carrier signal plus an infinite number of side frequency components [see (9.17)].

EXAMPLE 9-2

Suppose that a phase-modulated signal is given by

$$x(t) = 5\cos\left[3\pi(10^6)t + \frac{1}{4}\sin(100\pi t)\right]$$

What is the carrier frequency? If $k_p = 5$, find the message signal $m(t)$.

SOLUTION 9-2

The general form of a phase-modulated signal is given in (9.8), which we reproduce here for convenience:

$$x_{PM}(t) = A_c \cos[\omega_c t + k_p m(t)]$$

Comparison with the signal given in the problem description allows us to identify ω_c and $m(t)$. Notice that we can rewrite ω_c as follows:

$$\omega_c = 3\pi(10^6) = 2\pi\left(\frac{3}{2}\right)(10^6)$$

Therefore carrier frequency is $(3/2)(10^6) = 1.5$ MHz. Turning to the problem of identifying the message signal, let's put $k_p = 5$ into the general expression for a phase-modulated wave

$$x_{PM}(t) = A_c \cos[\omega_c t + k_p m(t)] = A_c \cos[\omega_c t + 5m(t)]$$

Therefore we have

$$5m(t) = \frac{1}{4}\sin(100\pi t) \Rightarrow m(t) = \frac{1}{20}\sin(100\pi t)$$

EXAMPLE 9-3
If an FM signal

$$x(t) = 5\cos\left[2\pi(10^6)t + \frac{1}{4}\sin(100\pi t)\right]$$

and $k_f = \pi$, identify the carrier frequency and the message signal.

SOLUTION 9-3
First we write down the general form of an FM signal as

$$x_{FM}(t) = A_c \cos\left[\omega_c t + k_f \int_0^t m(\tau)\,d\tau\right]$$

Comparison with

$$x(t) = 5\cos\left[2\pi(10^6)t + \frac{1}{4}\sin(100\pi t)\right]$$

tells us that the carrier radial frequency is given by $\omega_c = 2\pi(10^6)$. We conclude that the frequency is 10^6, or 1 MHz. To find the message signal, note that we have

$$k_f \int_0^t m(\tau)\, d\tau = \pi \int_0^t m(\tau)\, d\tau = \frac{1}{4}\sin(100\pi t)$$

Let's take the derivative of both sides. Using the fundamental theorem of calculus, we obtain

$$\frac{d}{dt}\left[\pi \int_0^t m(\tau)\, d\tau\right] = \pi m(t)$$

Taking the derivative of the right-hand side gives

$$\frac{d}{dt}\left(\frac{1}{4}\sin(100\pi t)\right) = 25\pi \cos(100\pi t)$$

Equating these terms gives us the message signal

$$m(t) = 25\cos(100\pi t)$$

EXAMPLE 9-4
Find the bandwidth of the angle-modulated signal

$$x(t) = 10\cos[2\pi(10^6)t + 50\sin(2000\pi t)]$$

SOLUTION 9-4
First we write down some key components of this signal. We have

$$\omega_c = 2\pi(10^6)$$

and

$$\phi(t) = 50\sin(2000\pi t)$$

Now we write down the instantaneous frequency, using (9.2):

$$\omega_i = \omega_c + \frac{d\phi}{dt} = 2\pi(10^6) + (10^5\pi)\cos(2000\pi t)$$

Next, we find $\Delta\omega = \max|\omega_i - \omega_c| = \max|(10^5\pi)\cos(2000\pi t)| = 10^5\pi$. We can use this to calculate the modulation index using $\beta = \Delta\omega/\omega_m$. The frequency of the message signal is $\omega_m = 2000\pi$ and so we have

$$\beta = \frac{10^5\pi}{2000\pi} = 50$$

The bandwidth can be calculated using (9.22) with β in place of D. We obtain

$$W_B = 2(\beta + 1)\omega_m = 2(51)(2000\pi) = 204\pi(10^3) \text{ rad/s}$$

EXAMPLE 9-5
If $\beta = 50$ and the frequency of the modulating signal is $f_m = 1$ kHz, is the signal narrowband or wideband?

SOLUTION 9-5
The signal is wideband; it does not satisfy $\beta \ll 1$.

EXAMPLE 9-6
Consider an FM signal. Suppose that $f_m = 120$ kHz and $\beta = 1$. What is the bandwidth of the signal?

SOLUTION 9-6
The bandwidth is given by $f_B = 2(\beta + 1)\,f_m = (4)(120)$ kHz $= 480$ kHz.

EXAMPLE 9-7
Consider an FM signal. Suppose that $f_m = 250$ kHz and $\beta = 0.15$. Find the bandwidth.

SOLUTION 9-7
The bandwidth is given by $f_B \approx 2f_m$ since $\beta \ll 1$. So we have $f_B = 2f_m = 2(250) = 500$ kHz.

Quiz

1. Consider a narrowband tone-modulated FM signal as given in (9.16). What is the main difference between this signal and an amplitude-modulated signal?
2. Let $x(t) = 2\cos[6\pi(10^6)t + \sin(1000\pi t)]$ be a phase-modulated signal. What is the frequency of the carrier?

3. If $x(t) = 2\cos[6\pi(10^6)t + \sin(1000\pi t)]$ is a phase-modulated signal and $k_p = 20$, what is the message signal?

4. Suppose that $x(t) = 2\cos[6\pi(10^6)t + t^2\sin(1000\pi t)]$ is an FM signal with $k_f = \pi$. What is the message signal?

5. Find the bandwidth of $x(t) = 5\cos[2\pi(10^6)t + 200\cos 2\pi 10^3 t]$.

6. Consider an FM signal. Suppose that $f_m = 550$ kHz and $\beta = 0.2$. Find the bandwidth.

CHAPTER 10

The Laplace Transform

The *Laplace transform* is a useful mathematical tool that converts continuous time-domain signals into a function of a complex variable denoted by s. This technique is very useful in the study of linear time-invariant (LTI) systems. Specifically, the Laplace transform allows us to convert ordinary differential equations into algebraic equations that are usually easier to solve. Moreover, the Laplace transform converts convolution into a simple multiplication, making it especially useful for the study of LTI systems since the output of an LTI system is given by the convolution of an input signal with the system's impulse-response function. In short, this is a tool that makes the study of continuous time LTI systems much easier.

Integral Formula and Definition

In the study of LTI systems we will use the Laplace transform to transform a continuous function of time t into a function of s, which is in general a complex number. We can write this new variable as follows:

$$s = \sigma + j\omega \tag{10.1}$$

where $\sigma = \text{Re}(s)$ and $\omega = \text{Im}(s)$ are real variables. The definition of the Laplace transform is given in terms of an integral. The Laplace transform $X(s)$ of a signal $x(t)$ is given by

$$X(s) = \int_{-\infty}^{\infty} x(t)\,e^{-st}\,dt \tag{10.2}$$

We can write this relation more abstractly as

$$X(s) = L\{x(t)\} \tag{10.3}$$

where $L\{\bullet\}$ is the Laplace transform viewed as an *operator* acting on the function $x(t)$. We will begin by showing how to directly compute the Laplace transform of a few functions using (10.2), but for the most part, the Laplace transform of common signals is given. (We list the most important Laplace transforms in Appendix B.) Generally speaking, it is important to learn how to manipulate Laplace transforms rather than calculate them. This process can be specified as follows:

- Given one or more input signals, look up their Laplace transforms in a table.
- Use the properties of the Laplace transform to accomplish various tasks *algebraically*. This will include solving differential equations or computing the convolution of two signals.
- A function of s will be the result. Manipulate this function until it is in a form that can be readily transformed back into a function of time by inspection.

In any case, it is a good idea to understand how to compute the Laplace transform directly before moving on, so we proceed with a few examples.

EXAMPLE 10-1
Find the Laplace transform of $x(t) = u(t)$ (see Fig. 10-1).

SOLUTION 10-1
Inserting this function into the defining integral (10.2), we find

$$X(s) = \int_{-\infty}^{\infty} u(t)\,e^{-st}\,dt = \int_{0}^{\infty} e^{-st}\,dt = -\frac{1}{s}e^{-st}\Big|_{0}^{\infty} = \frac{1}{s}$$

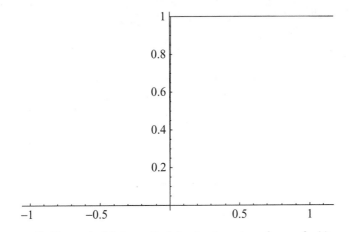

Fig. 10-1. In Example 10-1, we find the Laplace transform of $x(t) = u(t)$.

Therefore we have the Laplace transform pair

$$L\{u(t)\} = \frac{1}{s}$$

EXAMPLE 10-2
Find the Laplace transform of (see Fig. 10-2)

$$x(t) = e^{-at}u(t)$$

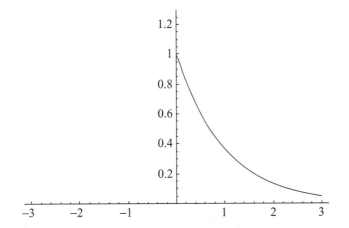

Fig. 10-2. In Example 10-2, we compute the Laplace transform of $x(t) = e^{-at}u(t)$.

SOLUTION 10-2
Again using (10.2), we have

$$X(s) = \int_{-\infty}^{\infty} e^{-at}u(t)\,e^{-st}\,dt = \int_{0}^{\infty} e^{-at}\,e^{-st}\,dt$$

$$= \int_{0}^{\infty} e^{-(a+s)t}\,dt = -\frac{1}{(s+a)}e^{-(s+a)t}\Big|_{0}^{\infty} = \frac{1}{(s+a)}$$

and so we have the Laplace transform pair

$$L\{e^{-at}u(t)\} = \frac{1}{s+a}$$

EXAMPLE 10-3
Find the Laplace transform of

$$x(t) = te^{-at}u(t)$$

SOLUTION 10-3
The function is displayed in Fig. 10-3. Proceeding, we have

$$X(s) = \int_{-\infty}^{\infty} te^{-at}u(t)\,e^{-st}\,dt = \int_{0}^{\infty} te^{-at}\,e^{-st}\,dt = \int_{0}^{\infty} te^{-(a+s)t}\,dt$$

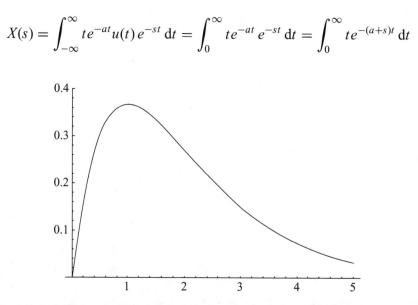

Fig. 10-3. In Example 10-3, we find the Laplace transform of $x(t) = te^{-at}u(t)$.

This is a familiar integral to most readers. No doubt many recall that this integral can be done using integration by parts. It is always good to review, and so let's quickly go through the process to refresh our memories. We start by recalling the integration by parts formula

$$\int u\,dv = uv - \int v\,du$$

In the case at hand, we set

$$u = t, \implies du = dt$$
$$dv = e^{-(s+a)t}\,dt, \implies v = -\frac{1}{s+a}e^{-(s+a)t}$$

Therefore, applying the integration by parts formula, we obtain

$$X(s) = -\frac{1}{s+a}te^{-(s+a)t}\Big|_0^\infty + \frac{1}{s+a}\int_0^\infty e^{-(s+a)t}\,dt$$

$$= \frac{1}{s+a}\int_0^\infty e^{-(s+a)t}\,dt$$

$$= -\frac{1}{(s+a)^2}e^{-(s+a)t}\Big|_0^\infty$$

$$= \frac{1}{(s+a)^2}$$

So we have derived the Laplace transform pair as

$$L\{te^{-at}u(t)\} = \frac{1}{(s+a)^2}$$

Important Properties of the Laplace Transform

The Laplace transform is a *linear* operation. This follows readily from the defining integral. Suppose that $X_1(s) = L\{x_1(t)\}$ and $X_2(s) = L\{x_2(t)\}$. Also, let α and β be two constants. Then

$$L\{\alpha x_1(t) + \beta x_2(t)\} = \int_{-\infty}^{\infty} (\alpha x_1(t) + \beta x_2(t))\, e^{-st}\, dt$$

Now we can just use the linearity properties of the integral. We have

$$\int_{-\infty}^{\infty} (\alpha x_1(t) + \beta x_2(t))\, e^{-st}\, dt = \int_{-\infty}^{\infty} \left(\alpha x_1(t) e^{-st} + \beta x_2(t) e^{-st}\right) dt$$

$$= \int_{-\infty}^{\infty} \alpha x_1(t) e^{-st}\, dt + \int_{-\infty}^{\infty} \beta x_2(t) e^{-st}\, dt$$

$$= \alpha \int_{-\infty}^{\infty} x_1(t) e^{-st}\, dt + \beta \int_{-\infty}^{\infty} x_2(t) e^{-st}\, dt$$

$$= \alpha X_1(s) + \beta X_2(s)$$

The next property we want to look at is *time scaling*. Time scaling in the case of the Laplace transform works just like it does in the case of the Fourier transform. Suppose that we have a continuous function of time $x(t)$ and some constant $a > 0$. Given that

$$X(s) = \int_{-\infty}^{\infty} x(t) e^{-st}\, dt$$

what is the Laplace transform of the time-scaled function $x(at)$? The defining integral is

$$L\{x(at)\} = \int_{-\infty}^{\infty} x(at)\, e^{-st}\, dt$$

Let's fix this up with a simple change of variables. Let

$$u = at, \;\Rightarrow\; du = a\, dt$$

Furthermore, we can write

$$t = \frac{u}{a}$$

and so we have

$$L\{x(at)\} = \int_{-\infty}^{\infty} x(at) e^{-st}\, dt = \frac{1}{a} \int_{-\infty}^{\infty} x(u) e^{-su/a}\, du$$

Let's write the argument of the exponential in a more suggestive way as

$$-\frac{su}{a} = -\left(\frac{s}{a}\right)u = -\theta u$$

where for the moment we have defined another new variable $\theta = s/a$. This change makes the above integral look just like a plain old Laplace transform. Since the integration variable u is just a "dummy" variable, we can call it anything we like. Let's put in the aforementioned changes and also let $u \rightarrow t$.

$$L\{x(at)\} = \frac{1}{a}\int_{-\infty}^{\infty} x(u)\,e^{-su/a}\,du$$

$$= \frac{1}{a}\int_{-\infty}^{\infty} x(u)\,e^{-\theta u}\,du$$

$$= \frac{1}{a}\int_{-\infty}^{\infty} x(t)\,e^{-\theta t}\,du$$

$$= \frac{1}{a}X(\theta)$$

Now we change back $\theta = s/a$, and we have discovered that

$$L\{x(at)\} = \frac{1}{a}X\left(\frac{s}{a}\right)$$

More generally, if we let a assume negative values as well, this relation is written as

$$L\{x(at)\} = \frac{1}{|a|}X\left(\frac{s}{a}\right) \tag{10.4}$$

The next property we wish to consider is *time shifting*. Supposing that

$$X(s) = L\{x(t)\}$$

what is the Laplace transform of $x(t - t_0)$? Using the definition (10.2), we have

$$L\{x(t - t_0)\} = \int_{-\infty}^{\infty} x(t - t_0)\,e^{-st}\,dt$$

Once again, we can proceed with a simple change of variables. We let $u = t - t_0$ from which it follows immediately that $du = dt$. Then

$$L\{x(t - t_0)\} = \int_{-\infty}^{\infty} x(u)\, e^{-s(u+t_0)}\, du$$

$$= \int_{-\infty}^{\infty} x(u)\, e^{-su}\, e^{-st_0}\, du$$

Notice that s and t_0 are not integration variables, and so given any term that is a function of these variables alone, we can just pull it outside the integral. This gives

$$L\{x(t - t_0)\} = e^{-st_0} \int_{-\infty}^{\infty} x(u)\, e^{-su}\, du = e^{-st_0} \int_{-\infty}^{\infty} x(t)\, e^{-st}\, dt = e^{-st_0} X(s)$$

We conclude that the effect of a time shift by t_0 is to multiply the Laplace transform by e^{-st_0}.

Differentiation

When we consider the derivative of a function in the time domain, we encounter one of the most useful properties of the Laplace transform which makes it well suited to use when solving ordinary differential equations. Starting with the defining integral, consider the Laplace transform

$$L\left\{\frac{dx}{dt}\right\}$$

We consider only those functions that vanish when $t < 0$. Using (10.2), this is just

$$\int_{0}^{\infty} \frac{dx}{dt}\, e^{-st}\, dt$$

Let's use our old friend integration by parts to move the derivative away from $x(t)$. We obtain

$$\int_{0}^{\infty} \frac{dx}{dt}\, e^{-st}\, dt = x(t)\, e^{-st}\, \Big|_{0}^{\infty} + s \int_{0}^{\infty} x(t)\, e^{-st}\, dt$$

We consider the boundary term first. Clearly $x(t)e^{-st}$ goes to zero at the upper limit because the decaying exponential goes to zero as $t \to \infty$. Therefore,

$$x(t)e^{-st}\Big|_0^\infty = -x(0)$$

Now take a look at the integral in the second term. This is nothing other than the Laplace transform of $x(t)$. So, we find that

$$L\left\{\frac{dx}{dt}\right\} = -x(0) + sX(s)$$

Now let's consider differentiation with respect to s; that is,

$$\frac{d}{ds}[X(s)] = \frac{d}{ds}\int_{-\infty}^\infty x(t)e^{-st}\,dt$$

The integration is with respect to t, so it seems fair enough that we can slide the derivative with respect to s on inside the integral. This gives

$$\frac{d}{ds}[X(s)] = \int_{-\infty}^\infty \frac{d}{ds}\left[x(t)e^{-st}\right]dt$$

$$= \int_{-\infty}^\infty x(t)\frac{d}{ds}\left(e^{-st}\right)dt = -\int_{-\infty}^\infty tx(t)e^{-st}\,dt$$

We can move the minus sign to the other side, i.e.,

$$-\frac{dX}{ds} = \int_{-\infty}^\infty tx(t)e^{-st}\,dt$$

This tells us that the Laplace transform of $tx(t)$ is given by

$$L\{tx(t)\} = -\frac{dX}{ds} \tag{10.5}$$

where $X(s)$ is the Laplace transform of $x(t)$. In the next two examples, we show how this can be useful.

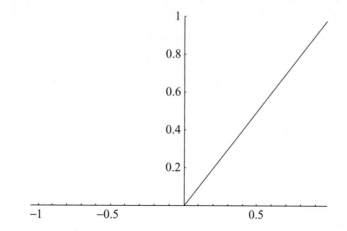

Fig. 10-4. In Example 10-4, we find the Laplace transform of $x(t) = tu(t)$.

EXAMPLE 10-4
Using the result of Example 10-1 and (10.5), find the Laplace transform of $x(t) = tu(t)$ (see Fig. 10-4).

SOLUTION 10-4
In Example 10-1 we showed that the Laplace transform of $u(t)$ was given by $1/s$. From (10.5), we know that the Laplace transform of $x(t) = tu(t)$ is

$$\text{L}\{tu(t)\} = -\frac{\text{d}}{\text{d}s}\left(\frac{1}{s}\right) = \frac{1}{s^2}$$

EXAMPLE 10-5
Given that

$$\text{L}\{\cos \beta tu(t)\} = \frac{s}{s^2 + \beta^2}$$

find $\text{L}\{t \cos \beta tu(t)\}$ (see Fig. 10-5).

SOLUTION 10-5
We obtain the result by computing the derivative of $s/(s^2 + \beta^2)$ and adding a minus sign. First we recall that the derivative of a quotient is given by

$$\left(\frac{f}{g}\right)' = \frac{f'g - g'f}{g^2}$$

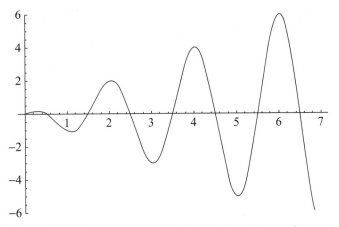

Fig. 10-5. In Example 10-5, we compute the Laplace transform of $t \cos \beta t u(t)$. The function is shown here with $\beta = \pi$.

In the case of $s/(s^2 + \beta^2)$, we have

$$f = s, \Rightarrow f' = 1$$
$$g = s^2 + \beta^2, \Rightarrow g' = 2s$$

and so,

$$\left(\frac{f}{g}\right)' = \frac{f'g - g'f}{g^2} = \frac{(s^2 + \beta^2) - 2s(s)}{(s^2 + \beta^2)^2} = \frac{\beta^2 - s^2}{(s^2 + \beta^2)^2}$$

Applying (10.5), we add a minus sign and find that

$$L\{t \cos \beta t u(t)\} = \frac{s^2 - \beta^2}{(s^2 + \beta^2)^2}$$

EXAMPLE 10-6
Find the solution of

$$\frac{dy}{dt} = A \cos t$$

for $t \geq 0$, where A is a constant and $y(0) = 1$.

SOLUTION 10-6
This is a very simple ordinary differential equation (ODE) and it can be verified by integration—that the solution is $y(t) = 1 + A \sin t$ (see Fig. 10-6). Since this

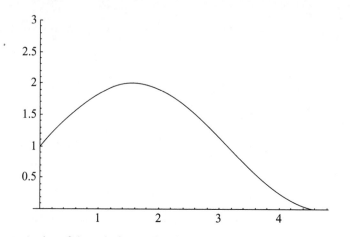

Fig. 10-6. A plot of the solution to $dy/dt = A \cos t$ with $A = 1$ and $y(0) = 1$.

is an easy ODE to solve, it is a good one to use to illustrate the method of the Laplace transform. Taking the Laplace transform of the left side, we have

$$L\left\{\frac{dy}{dt}\right\} = -y(0) + s\,Y(s) = -1 + s\,Y(s)$$

In the *Quiz*, we will show that

$$L\{\cos \beta t u(t)\} = \frac{s}{s^2 + \beta^2}$$

This tells us that the Laplace transform of the right-hand side of the differential equation is

$$L\{A \cos t\} = A\frac{s}{s^2 + 1}$$

(Remember, $t \geq 0$ was specified in the problem, and so we don't need to explicitly include the unit step function). Equating both sides gives us an equation we can solve algebraically as

$$-1 + s\,Y(s) = A\frac{s}{s^2 + 1}$$

Adding 1 to both sides, we obtain

$$sY(s) = 1 + A\frac{s}{s^2 + 1}$$

Now we divide through by s, giving an expression for the Laplace transform of $y(t)$:

$$Y(s) = \frac{1}{s} + A\frac{1}{s^2 + 1}$$

In Example 10-1, we found that

$$L\{u(t)\} = \frac{1}{s}$$

Since it has been specified that $t \geq 0$, this is the same as stating that

$$L\{1\} = \frac{1}{s}$$

In the *Quiz*, we will show that

$$X(s) = \frac{\beta}{s^2 + \beta^2}$$

is the Laplace transform of $x(t) = \sin(\beta t)\,u(t)$. Putting these results together, by inspection of

$$Y(s) = \frac{1}{s} + A\frac{1}{s^2 + 1}$$

we conclude that

$$y(t) = 1 + A\,\sin t$$

The Inverse Laplace Transform

The last example illustrates that we also need to go in the reverse direction; that is, given a function of s, we are going to find its representation as a function of t. This type of operation is known as the *inverse Laplace transform*. This is abstractly written as

$$x(t) = L^{-1}\{X(s)\} \tag{10.6}$$

Formally, the inverse Laplace transform can be calculated using the integral

$$x(t) = \frac{1}{2\pi j} \int_{c-j\infty}^{c+j\infty} X(s)\, e^{st}\, ds \tag{10.7}$$

This is an integral over the complex s plane. We are not going to worry about using this integral, in practice the inverse Laplace transform is calculated using a bit of intuition to manipulate expressions into recognizable terms. Frequently, this will involve the use of partial fraction decomposition, and more often than not we will use an intuitive approach to find the Laplace transform (or better yet—your favorite computer program).

Formally, we suppose that $X(s)$ can be written in the form

$$X(s) = \frac{A(s)}{B(s)} = k\frac{(s-z_1)(s-z_2)\cdots(s-z_m)}{(s-p_1)(s-p_2)\cdots(s-p_n)}$$

The *poles* of $X(s)$ are the complex numbers p_i, such that the denominator of the above expression $B(s) = 0$. If all the p_i are distinct and $m < n$, then we can put $X(s)$ into the form

$$X(s) = \frac{c_1}{s-p_1} + \cdots + \frac{c_n}{s-p_n} \tag{10.8}$$

where the c_i's are constants that can be calculated using

$$c_k = (s-p_k)\,X(s)\,|_{s=p_k} \tag{10.9}$$

On the other hand if $B(s)$ has factors that assume the form $(s-p_i)^r$, then we say that p_i is a pole of $X(s)$ with multiplicity r. In that case, we will be able to write $X(s)$ such that it consists of terms that look like

$$\frac{c_1}{s-p_i} + \frac{c_2}{(s-p_i)^2} + \cdots + \frac{c_r}{(s-p_i)^r} \tag{10.10}$$

In this case the constants are calculated using

$$c_{r-k} = \frac{1}{k!} \frac{d^k}{ds^k} \left[(s - p_i)^r X(s)\right]\Big|_{s=p_k} \qquad (10.11)$$

Finally, for the case where $m \geq n$, we can write $X(s)$ in the form

$$X(s) = Q(s) + \frac{R(s)}{D(s)}$$

and then compute the inverse transform. While these formulas provide a systematic framework for finding the inverse Laplace transform, we typically use an intuitive approach and tables to find them.

SHIFTING IN s

When trying to find inverse Laplace transforms, it is sometimes helpful to consider what happens when we shift in s. To find the inverse Laplace transform of $X(s - s_0)$, consider what happens when we compute the Laplace transform of $e^{s_0 t} x(t)$. Using (10.2), we find that

$$\int_{-\infty}^{\infty} e^{s_0 t} x(t) e^{-st} \, dt = \int_{-\infty}^{\infty} x(t) e^{-(s-s_0)t} \, dt = X(s - s_0)$$

So, if you see an expression that would look like the Laplace transform of something familiar but with $s \to s - s_0$, then the inverse Laplace transform is found from $e^{s_0 t} x(t)$ where $x(t)$ is the inverse Laplace transform of $X(s)$.

EXAMPLE 10-7
Find the inverse Laplace transform of

$$X(s) = \frac{s}{s^2 + 6s + 18} + \frac{3}{s^2 + 6s + 18}$$

SOLUTION 10-7
First we write this as a single term

$$X(s) = \frac{s + 3}{s^2 + 6s + 18}$$

The presence of $s + 3$ in the numerator is suggestive that this expression is the Laplace transform shifted by -3. So, we should try to write this expression

with an $s + 3$ in the denominator as well. Notice that

$$(s + 3)^2 + 9 = s^2 + 6s + 18$$

and so we have

$$X(s) = \frac{s + 3}{(s + 3)^2 + 9} \qquad (10.12)$$

Now if we had

$$\frac{s}{s^2 + 9}$$

then we know that the inverse Laplace transform would be

$$\cos 3t u(t)$$

So we see that, with $s + 3$ in place of s in (10.12), what we have is a shift in s by -3. So the inverse Laplace transform is

$$x(t) = e^{-3t} \cos 3t u(t)$$

EXAMPLE 10-8
Find the inverse Laplace transform of

$$X(s) = \frac{3s + 6}{s^2 + s - 6}$$

SOLUTION 10-8
First we factorize the denominator as

$$X(s) = \frac{3s + 6}{s^2 + s - 6} = \frac{3s + 6}{(s + 3)(s - 2)}$$

Now we use partial fraction decomposition to write

$$\frac{3s + 6}{(s + 3)(s - 2)} = \frac{A}{s + 3} + \frac{B}{s - 2} \qquad (10.13)$$

Let's cross multiply by $s - 2$ to obtain

$$\frac{3s + 6}{(s + 3)} = \frac{A(s - 2)}{s + 3} + B$$

Now set $s = 2$. The term involving A vanishes and we have

$$\frac{12}{5} = B$$

Now returning to (10.13), this time we cross multiply by $s + 3$ and obtain

$$\frac{3s + 6}{(s - 2)} = A + \frac{B(s + 3)}{s - 2}$$

Setting $s = -3$, this becomes

$$\frac{3s + 6}{(s - 2)} \rightarrow \frac{3}{5} = A$$

Therefore, we have

$$X(s) = \frac{3s + 6}{(s + 3)(s - 2)} = \frac{A}{s + 3} + \frac{B}{s - 2} = \left(\frac{3}{5}\right)\frac{1}{s + 3} + \left(\frac{12}{5}\right)\frac{1}{s - 2}$$

The expression is now in a form that can be inverted easily. Recalling Example 10-1, we see that

$$\frac{1}{s + 3} \rightarrow e^{-3t}u(t)$$

For the second term, we obtain

$$\frac{1}{s - 2} \rightarrow e^{2t}u(t)$$

Putting everything together, the inverse Laplace transform of $X(s)$ is

$$x(t) = \frac{3}{5}e^{-3t}u(t) + \frac{12}{5}e^{2t}u(t)$$

Linear Systems and Convolution

Consider a continuous time LTI system. Recall that the response $y(t)$ to any input $x(t)$ is given in terms of the impulse response $h(t)$ as

$$y(t) = x(t)*h(t) \qquad\qquad (10.14)$$

Since the Laplace transform converts convolution into simple multiplication, if we can find $Y(s)$, $X(s)$, and $H(s)$ then

$$Y(s) = X(s)H(s) \qquad\qquad (10.15)$$

The Laplace transform of the impulse response function, $H(s)$, is known as the *transfer function*. Using (10.15), if we know the output response for a given input, then we can find the transfer function of the system readily:

$$H(s) = \frac{Y(s)}{X(s)}$$

We can learn some basic characteristics of the system by examining the transfer function. If we consider systems for which we ignore $x(t)$ for $t < 0$, then the transfer function is defined for *relaxed* systems. In this case all initial conditions are set to zero. We assume that this is the case in the following examples.

EXAMPLE 10-9
Consider an LTI system for which

$$\frac{dy}{dt} = \frac{dx}{dt} - 3x(t)$$

Assume that the system is causal and find the impulse response $h(t)$.

SOLUTION 10-9
We take the Laplace transform, giving

$$sY(s) = sX(s) - 3X(s)$$

Some manipulation leads to

$$sY(s) = X(s)(s - 3)$$

The transfer function is

$$H(s) = \frac{Y(s)}{X(s)} = \frac{s-3}{s} = 1 - \frac{3}{s}$$

Taking the inverse Laplace transform gives the impulse response function

$$h(t) = \delta(t) - 3u(t)$$

EXAMPLE 10-10
Consider an LTI system for which

$$\frac{dy}{dt} + 4y(t) = \frac{dx}{dt} + x(t)$$

Assume that the system is causal and find the impulse response $h(t)$.

SOLUTION 10-10
We take the Laplace transform, giving

$$sY(s) + 4Y(s) = sX(s) + X(s)$$

Some algebra leads to

$$Y(s)(s+4) = X(s)(s+1)$$

Solving we find the transfer function

$$H(s) = \frac{Y(s)}{X(s)} = \frac{s}{s+4} + \frac{1}{s+4}$$

To find the impulse response function, we take the inverse Laplace transform of this expression. The second term is easy. We have the following Laplace transform pair:

$$\frac{1}{s+4} \leftrightarrow e^{-4t}u(t)$$

The first term cannot be inverted by inspection. But let's rewrite it so that we can arrive at terms that can be inverted immediately. We have

$$\frac{s}{s+4} = \frac{s+4-4}{s+4} = \frac{s+4}{s+4} - \frac{4}{s+4} = 1 - \frac{4}{s+4}$$

Now we can find the impulse response function. We have

$$H(s) = \frac{s}{s+4} + \frac{1}{s+4} = 1 - \frac{4}{s+4} + \frac{1}{s+4} = 1 - \frac{3}{s+4}$$

and so we conclude that

$$h(t) = \delta(t) - 3\,e^{-4t}u(t)$$

LTI Systems in Series and Parallel

Consider that two LTI systems are in series (see Fig. 10-7). What happens if we consider this to be a single system, that is only worry about the input to system 1 and the output of system 2? In the case of LTI systems in series, the combined transfer function is the product of the individual transfer functions; that is,

$$H(s) = H_1(s)H_2(s) \qquad (10.16)$$

For two systems in parallel (see Fig. 10-8), the transfer function of the composite system is given by the *sum* of the individual transfer functions; that is,

$$H(s) = H_1(s) + H_2(s) \qquad (10.17)$$

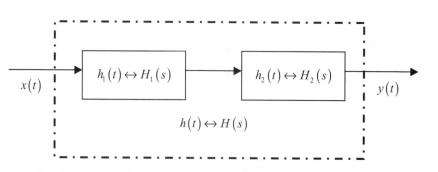

Fig. 10-7. Two LTI systems in series. We can view them as a single system with transfer function given by (10.16).

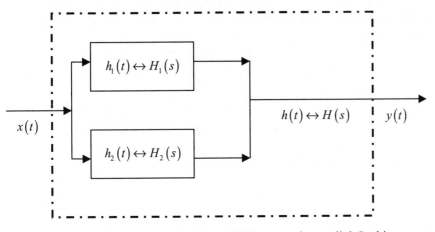

Fig. 10-8. A schematic representation of two LTI systems in parallel. In this case, we sum the transfer functions of the individual systems to obtain the overall transfer function, as described by (10.17).

EXAMPLE 10-11

Two systems are arranged in series with

$$h_1(t) = e^{-2t}u(t)$$
$$h_2(t) = e^{-4t}u(t)$$

Find the impulse response of the entire system.

SOLUTION 10-11

First we find the Laplace transform of each function as

$$H_1(s) = \frac{1}{s+2}, \qquad H_2(s) = \frac{1}{s+4}$$

Since the systems are arranged in series, the transfer function for the entire system is the product of these two transfer functions; that is,

$$H(s) = H_1(s)H_2(s) = \left(\frac{1}{s+2}\right)\left(\frac{1}{s+4}\right)$$

We do a partial fraction decomposition and write

$$\left(\frac{1}{s+2}\right)\left(\frac{1}{s+4}\right) = \frac{A}{s+2} + \frac{B}{s+4}$$

Now we cross multiply to obtain

$$1 = A(s + 4) + B(s + 2)$$

Setting $s = -2$ leads to

$$1 = 2A, \implies A = 1/2$$

Setting $s = -4$ leads to

$$1 = -2B, \implies B = -1/2$$

and so we can write the transfer function for the system as

$$H(s) = \left(\frac{1}{2}\right)\frac{1}{s+2} - \left(\frac{1}{2}\right)\frac{1}{s+4}$$

Taking the inverse Laplace transform gives the impulse response

$$h(t) = \frac{1}{2}u(t)\left(e^{-2t} - e^{-4t}\right)$$

EXAMPLE 10-12
Find the convolution

$$e^{2t} * e^{-t}$$

in two ways.

SOLUTION 10-12
First we do it by integration. We have

$$e^{2t} * e^{-t} = \int_0^t e^{2x}\, e^{-(t-x)}\, dx = e^{-t}\int_0^t e^{3x}\, dx = \frac{1}{3}e^{-t}\, e^{3x}\,\Big|_0^t$$

$$= \frac{1}{3}e^{-t}(e^{3t} - 1) = \frac{1}{3}(e^{2t} - e^{-t})$$

Now, using the Laplace transform, the convolution $f(t)*g(t)$ is given by $F(s)G(s)$. The Laplace transform of e^{2t} is given by $1/(s - 2)$ and the Laplace transform of e^{-t} is given by $1/(s-1)$. The product is

$$\frac{1}{s-2}\frac{1}{s+1} = \frac{A}{s-2} + \frac{B}{s+1}$$

where we have gone forward using partial fraction decomposition. Now we cross multiply to obtain

$$1 = A(s+1) + B(s-2)$$

Next we set $s = 2$ to eliminate B. This gives

$$A = \frac{1}{3}$$

Setting $s = -1$ to eliminate A gives $B = -1/3$. So we arrive at

$$\frac{1}{s-2}\frac{1}{s+1} = \left(\frac{1}{3}\right)\frac{1}{s-2} - \left(\frac{1}{3}\right)\frac{1}{s+1}$$

Taking the inverse Laplace transform, we obtain

$$e^{2t} * e^{-t} = \frac{1}{3}\left(e^{2t} - e^{-t}\right)$$

Second-Order Systems

We have seen in the examples that the Laplace transform is useful for solving first-order systems. In many cases a system will include a second-order derivative. The Laplace transform is very useful in these situations as well, turning a differential equation into an algebraic one—although in this case we have one higher order term plus an extra term for the initial conditions (as you would expect). The Laplace transform pair of a second derivative is given by

$$\frac{d^2x}{dt^2} \rightleftharpoons s^2 X(s) - sx(0) - x'(0) \tag{10.18}$$

To see how to solve a second-order system, we consider an example.

EXAMPLE 10-13

Consider a continuous time LTI system described by

$$y''(t) + y'(t) - 6y(t) = x(t)$$

If the system is causal, find the impulse response $h(t)$.

SOLUTION 10-13

We compute the Laplace transform to find

$$s^2 Y(s) + s Y(s) - 6Y(s) = X(s)$$

Factoring the common term $Y(s)$ on the left-hand side, we have

$$\left(s^2 + s - 6\right) Y(s) = X(s)$$

Therefore the system function is given by

$$H(s) = \frac{Y(s)}{X(s)} = \frac{1}{s^2 + s - 6}$$

The right-hand side is easily factored, giving

$$H(s) = \frac{1}{(s + 3)(s - 2)}$$

The partial fraction expansion of this expression is

$$H(s) = \frac{1}{(s + 3)(s - 2)} = \frac{A}{s + 3} + \frac{B}{s - 2}$$

Let's cross multiply by $s + 3$, giving

$$\frac{1}{(s - 2)} = A + \frac{B}{s - 2}(s + 3)$$

Setting $s = -3$ gives $A = -1/5$. Using a similar procedure to solve for B, we find

$$H(s) = -\frac{1}{5}\frac{1}{s + 3} + \frac{1}{5}\frac{1}{s - 2}$$

Application of the inverse Laplace transform gives us the impulse response

$$h(t) = -\frac{1}{5} \left(e^{-3t} - e^{2t} \right) u(t)$$

Quiz

Find the Laplace transforms of the following signals:

1. $x(t) = \delta(t)$.
2. $x(t) = \cos(\beta t) u(t)$.
3. $x(t) = \sin(\beta t) u(t)$.
4. Given your result from the previous problem, find the Laplace transform of $x(t) = t \sin(\beta t) u(t)$. Assume that $\beta > 0$ is a real constant.
5. Using Laplace transform methods, solve

$$\frac{dy}{dt} + 2y(t) = e^{-t}$$

   where $y(0)$ and $t \geq 0$.
6. Find the inverse Laplace transform of

$$X(s) = \frac{7s + 9}{s^2 - s - 2}$$

7. Find the inverse Laplace transform of $X(s) = e^{-as}$.
8. Use Laplace transform methods to find the impulse response for a system described by

$$\frac{dy}{dt} + 2y(t) = 3\frac{dx}{dt} + 2x(t)$$

9. Two systems are arranged in series with

$$h_1(t) = 4\delta(t)$$
$$h_2(t) = e^{-4t} u(t)$$

   Find the impulse response of the entire system.
10. Find $x(t) * \delta(t)$.

CHAPTER 11

The *z*-Transform

The *z*-transform is a transform that is applied to discrete time signals. Here *z* is a complex variable. This transform simplifies the analysis of discrete time signals by allowing us to convert finite difference equations into algebraic equations. The *z*-transform of a discrete time signal $x[n]$, denoted by $X(z)$, is given by

$$X(z) = \sum_{n=-\infty}^{\infty} x[n]z^{-n} \qquad (11.1)$$

Given that *z* is a complex number, it can be written in polar representation. Typically, this is written as

$$z = re^{j\Omega} \qquad (11.2)$$

If a signal $x[n]$ is zero when $n < 0$, then we can calculate the *z*-transform using the unilateral formula:

$$X_I(z) = \sum_{n=0}^{\infty} x[n]z^{-n} \qquad (11.3)$$

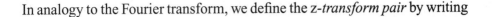

In analogy to the Fourier transform, we define the z-*transform pair* by writing

$$x[n] \leftrightarrow X(z) \qquad\qquad (11.4)$$

Basic Properties of the z-Transform

Let a and b be constants, defined as two discrete time signals $x_1[n]$ and $x_2[n]$. We consider the sum

$$y[n] = ax_1[n] + bx_2[n]$$

Using (11.1), the z-transform of $y[n]$ is found to be

$$Y(z) = \sum_{n=-\infty}^{\infty} y[n]z^{-n} = \sum_{n=-\infty}^{\infty} (ax_1[n] + bx_2[n])z^{-n}$$

$$= a \sum_{n=-\infty}^{\infty} x_1[n]z^{-n} + b \sum_{n=-\infty}^{\infty} x_2[n]z^{-n}$$

In other words, $Y(z) = aX_1(z) + bX_2(z)$. The z-transform is *linear*.

Next let's consider a time-shifted signal, which we denote by $x[n - n_0]$. The z-transform of this signal is

$$\sum_{n=-\infty}^{\infty} x[n - n_0]z^{-n}$$

Let $m = n - n_0$. Then $n = m + n_0$, and clearly as $n \to \pm\infty$, we have $m \to \pm\infty$. Then the z-transform of the time-shifted signal becomes

$$\sum_{n=-\infty}^{\infty} x[n - n_0]z^{-n} = \sum_{m=-\infty}^{\infty} x[m]z^{-(m+n_0)}$$

$$= z^{-n_0} \sum_{m=-\infty}^{\infty} x[m]z^{-m} = z^{-n_0} X(z)$$

So we conclude that a time shift by n_0 results in multiplication of the z-transform by z^{-n_0}. In other words, we have the z-transform pair

$$x[n - n_0] \leftrightarrow z^{-n_0} X(z) \tag{11.5}$$

Now let's consider *time reversal*; that is, what is the z-transform of $x[-n]$? We follow the same procedure that we used when considering time shifting. The z-transform is given by

$$\sum_{n=-\infty}^{\infty} x[-n]z^{-n}$$

Now let $m = -n$, when $n = -\infty$, $m = +\infty$ and when $n = +\infty$, $\in m = -\infty$. So we have

$$\sum_{n=-\infty}^{\infty} x[-n]z^{-n} = \sum_{m=\infty}^{-\infty} x[m]z^{m}$$

We're just adding up terms, so that we can swap these limits around.

$$\sum_{m=\infty}^{-\infty} x[m]z^{m} = \sum_{m=-\infty}^{\infty} x[m]z^{m} = \sum_{m=-\infty}^{\infty} x[m]\left(\frac{1}{z}\right)^{-m} = X\left(\frac{1}{z}\right)$$

This gives us another z-transform pair

$$x[-n] \leftrightarrow X\left(\frac{1}{z}\right) \tag{11.6}$$

Now considering differentiation of (11.1) with respect to z, we find that

$$\frac{dX}{dz} = \frac{d}{dz}\left[\sum_{n=-\infty}^{\infty} x[n]z^{-n}\right] = \sum_{n=-\infty}^{\infty} x[n]\frac{d}{dz}(z^{-n}) = \sum_{n=-\infty}^{\infty} -nx[n]z^{-n-1}$$

Now let's pull the minus sign outside of the summation as

$$\frac{dX}{dz} = -\sum_{n=-\infty}^{\infty} nx[n]z^{-n-1}$$

Next, note that we can write

$$z^{-n-1} = z^{-n}z^{-1} = \frac{z^{-n}}{z}$$

and so we obtain

$$\frac{dX}{dz} = -\sum_{n=-\infty}^{\infty} nx[n]\left(\frac{z^{-n}}{z}\right) = -\frac{1}{z}\sum_{n=-\infty}^{\infty} nx[n]z^{-n}$$

Multiplication of both side of z by $-z$ gives

$$-z\frac{dX}{dz} = \sum_{n=-\infty}^{\infty} nx[n]z^{-n}$$

This gives us another z-transform pair; multiplication of a discrete time signal by n results in differentiation of z:

$$nx[n] \leftrightarrow -z\frac{dX}{dz} \qquad (11.7)$$

Additional Properties of the z-Transform

We present three additional properties of the z-transform without proof. The first is a familiar one—convolution in the time domain is transformed into multiplication in the z domain; that is,

$$x_1[n]^*x_2[n] = X_1(z)X_2(z) \qquad (11.8)$$

Next, we consider the *accumulation property*. This involves the z-transform of the sum $\sum_{k=-\infty}^{\infty} x[k]$. We have the following z-transform pair:

$$\sum_{k=-\infty}^{\infty} x[k] \leftrightarrow \frac{1}{1-z^{-1}}X(z) \qquad (11.9)$$

The final property of the z-transform is multiplication of a discrete time signal by z_0^n. In this case we have

$$z_0^n x[n] \leftrightarrow X\left(\frac{z}{z_0}\right) \tag{11.10}$$

EXAMPLE 11-1
Find the z-transform of the unit impulse sequence $\delta[n]$.

SOLUTION 11-1
First, we write the definition of the z-transform as

$$X(z) = \sum_{n=-\infty}^{\infty} x[n]z^{-n}$$

In this case, we have

$$X(z) = \sum_{n=-\infty}^{\infty} \delta[n]z^{-n}$$

Now the unit impulse sequence is defined by

$$\delta(n) = \begin{cases} 1 & n = 0 \\ 0 & n \neq 0 \end{cases}$$

This tells us that only one term from the sum survives, namely the term with $n = 0$, and so

$$X(z) = \sum_{n=-\infty}^{\infty} \delta[n]z^{-n} = 1$$

The Region of Convergence (ROC)

The reader no doubt knows that when working with infinite series it is important to know when the series converges. In the case of the z-transform, the range of values of z for which the series converges is known as the *region of convergence* or ROC for short. This concept is typically illustrated by considering the unit step sequence. Let's recall the basic definition

$$u[n] = \begin{cases} 1 & n \geq 0 \\ 0 & n < 0 \end{cases}$$

Now we consider a real constant a, and then consider the z-transform of $a^n u[n]$. Looking at the definition of the unit step sequence given above, we see that we can use (11.3) and thus we have

$$X(z) = \sum_{n=-\infty}^{\infty} a^n u[n] z^{-n} = \sum_{n=0}^{\infty} a^n z^{-n} = \sum_{n=0}^{\infty} \left(\frac{a}{z}\right)^n \qquad (11.11)$$

Now the famous geometric series $\sum_{n=0}^{\infty} br^n$ is convergent if and only if $|r| < 1$, in which

$$\sum_{n=0}^{\infty} br^n = \frac{b}{1-r}$$

Comparison with (11.11) indicates that the z-transform of $a^n u[n]$ will be convergent if

$$\left|\frac{a}{z}\right| < 1$$

or if $|z| > |a|$. Using the formula for the geometric series, we set $b = 1$ and $r = a/z$ to give

$$X(z) = \frac{1}{1 - \frac{a}{z}} = \frac{1}{\frac{1}{z}(z-a)} = \frac{z}{z-a}$$

Again, this result is true provided that $|z| > |a|$, which defines the region of convergence.

EXAMPLE 11-2
Find the z-transform of $x[n] = \cos(\omega n) u[n]$ and discuss the region of convergence.

SOLUTION 11-2

Using (11.1), we have

$$X(z) = \sum_{n=-\infty}^{\infty} x[n]z^{-n} = \sum_{n=-\infty}^{\infty} \cos(\omega n)u[n]z^{-n}$$

Applying the definition of the unit step sequence $u[n]$, this becomes

$$X(z) = \sum_{n=0}^{\infty} \cos(\omega n)z^{-n}$$

First, we use Euler's identity to write the cosine function in terms of exponentials

$$\cos(\omega n) = \frac{e^{j\omega n} + e^{-j\omega n}}{2}$$

and so the summation becomes

$$X(z) = \sum_{n=0}^{\infty} \cos(\omega n)z^{-n} = \sum_{n=0}^{\infty} \left(\frac{e^{j\omega n} + e^{-j\omega n}}{2} \right) z^{-n}$$

$$= \frac{1}{2} \sum_{n=0}^{\infty} e^{j\omega n} z^{-n} + \frac{1}{2} \sum_{n=0}^{\infty} e^{-j\omega n} z^{-n} \qquad (11.12)$$

Now let's write z in polar form. Recalling (11.2), this is

$$z = re^{j\Omega}$$

Writing z in polar form allows us to write the first summation as

$$\frac{1}{2} \sum_{n=0}^{\infty} e^{j\omega n} z^{-n} = \frac{1}{2} \sum_{n=0}^{\infty} e^{j\omega n} (re^{j\Omega})^{-n}$$

$$= \frac{1}{2} \sum_{n=0}^{\infty} r^{-n} e^{j\omega n} e^{-j\Omega n} = \frac{1}{2} \sum_{n=0}^{\infty} r^{-n} e^{j(\omega - \Omega)n}$$

Similarly, the second summation in (11.12) can be written as

$$\frac{1}{2} \sum_{n=0}^{\infty} r^{-n} e^{-j(\omega+\Omega)n}$$

To find the region of convergence, we need to consider only one of these summations. Let's use the first. We can ignore the $\frac{1}{2}$ on the outside since it is not relevant to convergence. Rewriting slightly, we have

$$\sum_{n=0}^{\infty} r^{-n} e^{j(\omega-\Omega)n} = \sum_{n=0}^{\infty} \frac{e^{j(\omega-\Omega)n}}{r^n} = \sum_{n=0}^{\infty} \left(\frac{e^{j(\omega-\Omega)}}{r}\right)^n$$

Once again we have a geometric series. This series will converge provided that

$$\left|\frac{e^{j(\omega-\Omega)}}{r}\right| < 1 \qquad (11.13)$$

Now this is a complex number, and $|\alpha| = \sqrt{\alpha\alpha^*}$ for any complex number α. The complex conjugate of the term of interest is

$$\left(\frac{e^{j(\omega-\Omega)}}{r}\right)^* = \frac{e^{-j(\omega-\Omega)}}{r}$$

and so the magnitude is

$$\left|\frac{e^{j(\omega-\Omega)}}{r}\right| = \sqrt{\left(\frac{e^{j(\omega-\Omega)}}{r}\right)\left(\frac{e^{-j(\omega-\Omega)}}{r}\right)} = \sqrt{\frac{1}{r^2}} = \frac{1}{r}$$

But by definition, $z = r e^{j\Omega}$ where $|z| = r$. This means that the condition for convergence for the series given in (11.13) becomes

$$\frac{1}{|z|} < 1, \quad \text{i.e., } |z| > 1$$

Thus the region of convergence is $|z| > 1$. Now we can use the geometric series result to find the z-transform of $x[n] = \cos(\omega n)u[n]$. Let's call together our

Signals and Systems Demystified

results. In (11.12) we found that

$$X(z) = \frac{1}{2} \sum_{n=0}^{\infty} e^{j\omega n} z^{-n} + \frac{1}{2} \sum_{n=0}^{\infty} e^{-j\omega n} z^{-n}$$

Restating the general result for a geometric series,

$$\sum_{n=0}^{\infty} br^n = \frac{b}{1-r}$$

we see that the first summation is

$$\frac{1}{2} \sum_{n=0}^{\infty} e^{j\omega n} z^{-n} = \frac{1}{2} \sum_{n=0}^{\infty} \left(\frac{e^{j\omega}}{z}\right)^n = \frac{1}{2}\left(\frac{1}{1-e^{j\omega}/z}\right) = \frac{1}{2} \frac{z}{z - e^{j\omega}} \qquad (11.14)$$

Similarly, the second summation is

$$\frac{1}{2} \sum_{n=0}^{\infty} e^{-j\omega n} z^{-n} = \frac{1}{2} \frac{z}{z - e^{-j\omega}} \qquad (11.15)$$

Now we use (11.12) together with (11.14) and (11.15) to write

$$X(z) = \frac{1}{2}\left[\frac{z}{z - e^{j\omega}} + \frac{z}{z - e^{-j\omega}}\right]$$

We can manipulate this equation so that both terms have a common denominator. For the first term, we find that

$$\frac{z}{z - e^{j\omega}} = \frac{z}{z - e^{j\omega}}\left(\frac{z - e^{-j\omega}}{z - e^{-j\omega}}\right)$$

$$= \frac{z^2 - z e^{-j\omega}}{z^2 - z e^{j\omega} - z e^{-j\omega} + 1} = \frac{z^2 - z e^{-j\omega}}{z^2 - 2z \cos\omega + 1}$$

To obtain this result, we used Euler's identity

$$z e^{j\omega} + z e^{-j\omega} = z(e^{j\omega} + e^{-j\omega}) = 2z\left(\frac{e^{j\omega} + e^{-j\omega}}{2}\right) = 2z \cos\omega$$

For the second term in the z-transform, we find

$$\frac{z}{z - e^{-j\omega}} = \frac{z}{z - e^{-j\omega}}\left(\frac{z - e^{j\omega}}{z - e^{j\omega}}\right) = \frac{z^2 - z\,e^{j\omega}}{z^2 - 2z\,\cos\omega + 1}$$

Finally, putting everything together, we obtain

$$X(z) = \frac{1}{2}\left[\frac{z}{z - e^{j\omega}} + \frac{z}{z - e^{-j\omega}}\right] = \frac{z^2 - z\cos\omega}{z^2 - 2z\,\cos\omega + 1} \qquad (11.16)$$

This defines the z-transform pair

$$\cos(\omega n)u[n] \leftrightarrow \frac{z^2 - z\cos\omega}{z^2 - 2z\,\cos\omega + 1} \qquad (11.17)$$

provided that $|z| > 1$.

The Inverse z-Transform

We now consider how to proceed in the opposite direction; that is, given an expression written in terms of z, how we invert it to obtain a discrete time sequence $x[n]$. The first method we will investigate is the method of partial fractions. In the following examples, recall the z-transform pair

$$a^n u[n] \leftrightarrow \frac{z}{z - a} \qquad (11.18)$$

EXAMPLE 11-3
Find the inverse z-transform of

$$X(z) = \frac{z}{(z - 3)(z + 4)}$$

SOLUTION 11-3
Using partial fractions, we write

$$\frac{X(z)}{z} = \frac{1}{(z - 3)(z + 4)} = \frac{A}{z - 3} + \frac{B}{z + 4} \qquad (11.19)$$

The goal is to solve for the unknowns A and B. We begin by multiplying both sides by the product $(z - 3)(z + 4)$ as

$$1 = A(z + 4) + B(z - 3)$$

Notice that if we set $z = 3$ then the term involving B will vanish. Doing this, we obtain

$$1 = 7A \Rightarrow A = 1/7$$

Now we wish to eliminate the term involving A, and so we set $z = -4$ and find that

$$1 = -7B \Rightarrow B = -1/7$$

With these results in hand, (11.19) becomes

$$\frac{X(z)}{z} = \frac{1}{7}\left(\frac{1}{z - 3} - \frac{1}{z + 4}\right)$$

Multiplying this expression by z, we get

$$X(z) = \frac{1}{7}\left(\frac{z}{z - 3} - \frac{z}{z + 4}\right)$$

This expression can be inverted using (11.18). The answer is

$$x[n] = \frac{1}{7}3^n u[n] - \frac{1}{7}(-4)^n u[n] = \frac{1}{7}u[n][3^n - (-4)^n]$$

EXAMPLE 11-4
Find the inverse z-transform of

$$X(z) = \frac{z}{(z - 5)(z - 2)^2}$$

SOLUTION 11-4
The partial fraction expansion of this expression is

$$\frac{X(z)}{z} = \frac{1}{(z - 5)(z - 2)^2} = \frac{A}{z - 5} + \frac{B}{z - 2} + \frac{C}{(z - 2)^2}$$

Cross-multiplication leads to

$$1 = A(z-2)^2 + B(z-5)(z-2) + C(z-5)$$

First, we set $z = 5$ and obtain

$$1 = A(5-2)^2 = 9A \Rightarrow A = 1/9$$

Now we set $z = 2$ and find that

$$1 = C(2-5) = -3C \Rightarrow C = -1/3$$

Now we set $z = 0$ to find the last unknown, which is B. This gives

$$1 = 4/9 + 10B + 5/3 \Rightarrow B = -1/9$$

Therefore,

$$X(z) = \left(\frac{1}{9}\right)\frac{z}{z-5} - \left(\frac{1}{9}\right)\frac{z}{z-2} - \left(\frac{1}{3}\right)\frac{z}{(z-2)^2}$$

The region of convergence is $|z| > 5$. Using (11.18), we see that

$$\left(\frac{1}{9}\right)\frac{z}{z-5} \leftrightarrow \left(\frac{1}{9}\right)5^n u[n]$$

$$\left(\frac{1}{9}\right)\frac{z}{z-2} \leftrightarrow \left(\frac{1}{9}\right)2^n u[n]$$

To find the inverse of the last term, we apply the differentiation property (11.7) to (11.18). This gives

$$na^n u[n] \leftrightarrow \frac{az}{(z-a)^2}$$

with a region of convergence $|z| > |a|$. So the last term is inverted to give

$$\left(\frac{1}{3}\right)\frac{z}{(z-2)^2} \leftrightarrow \left(\frac{1}{3}\right)n2^n u[n]$$

Putting everything together, we obtain

$$x[n] = \frac{u[n]}{9}\{5^n - 2^n(1 + 3n)\}$$

Power Series Expansion

We can write the z-transform as a power series where the coefficients of each power of z^{-n} are the values of the sequence $x[n]$; that is,

$$X(z) = \cdots + x[-2]z^2 + x[-1]z + x[0] + x[1]z^{-1}$$
$$+ x[2]z^{-2} + x[3]z^{-3} + \cdots \tag{11.20}$$

If we consider only those sequences that are 0 for $n < 0$, then

$$X(z) = x[0] + x[1]z^{-1} + x[2]z^{-2} + x[3]z^{-3} + \cdots$$

This method is helpful for finding the z-transform of a sequence of numbers, as the next example shows.

EXAMPLE 11-5
Find the z-transform of the sequence $\{1, 2, 2, 4, 5, 1\}$. Assume that $x[n] = 0$ when $n < 0$.

SOLUTION 11-5
Applying (11.20), we find

$$X(z) = 1 + 2z^{-1} + 2z^{-2} + 4z^{-3} + 5z^{-4} + z^{-5}$$

EXAMPLE 11-6
Find the inverse z-transform of $X(z) = 1 + z^{-1} + 3z^{-2} - 2z^{-3}$.

SOLUTION 11-6
We simply apply (11.20) in reverse to obtain

$$x[n] = \{1, 1, 3, -2\}$$

Linear Time-Invariant Systems and the z-Transform

Recalling that the output of a linear time-invariant (LTI) system is given by

$$y[n] = x[n]^*h[n]$$

where $h[n]$ is the impulse response, we can use the convolution property of the z-transform (11.8) to write

$$Y(z) = X(z)H(z) \qquad (11.21)$$

Provided that we can find the z-transforms of all the signals, given that convolution is usually painful to compute, (11.21) can simplify the analysis of a discrete time LTI system. Notice that, given any input $X(z)$ and output $Y(z)$, we can use (11.21) to deduce the impulse response as

$$H(z) = \frac{Y(z)}{X(z)} \qquad (11.22)$$

EXAMPLE 11-7
Given that $x[n] = \{1, 2, 1, 2\}$ and $h[n] = \{1, 1, 1\}$, find the response $y[n]$.

SOLUTION 11-7
We could compute the convolution $y[n] = x[n]^*h[n]$ and give ourselves a headache, but instead we find the z-transforms and compute $Y(z) = X(z)H(z)$. We have

$$x[n] = \{1, 2, 1, 2\}$$

This tells us that

$$X(z) = 1 + 2z^{-1} + z^{-2} + 2z^{-3}$$

For the impulse response $h[n] = \{1, 1, 1\}$, we obtain

$$H(z) = 1 + z^{-1} + z^{-2}$$

The response is then

$$Y(z) = X(z)H(z)$$
$$= (1 + 2z^{-1} + z^{-2} + 2z^{-3})(1 + z^{-1} + z^{-2})$$
$$= 1 + 3z^{-1} + 4z^{-2} + 5z^{-3} + 3z^{-4} + 2z^{-5}$$

This can be inverted immediately to obtain

$$y[n] = \{1, 3, 4, 5, 3, 2\}$$

EXAMPLE 11-8

A causal discrete time LTI system is described by

$$y[n] - 2y[n-1] = x[n]$$

Find the impulse response of the system.

SOLUTION 11-8

Taking the z-transform of this difference equation, we obtain

$$Y(z) - 2z^{-1}Y(z) = X(z)$$

which leads to

$$Y(z)(1 - 2z^{-1}) = X(z)$$

We can immediately solve this to find $H(z)$ as

$$H(z) = \frac{Y(z)}{X(z)} = \frac{1}{1 - 2z^{-1}} = \frac{z}{z - 2}$$

Computing the inverse z-transform, we find that

$$h[n] = 2^n u[n]$$

EXAMPLE 11-9

A causal discrete time LTI system is described by

$$y[n] - 3y[n-1] + 2y[n-2] = x[n]$$

Find the impulse response of the system.

SOLUTION 11-9

Taking the *z*-transform

$$Y(z) - 3z^{-1}Y(z) + 2z^{-2}Y(z) = X(z)$$

leads to

$$Y(z)(1 - 3z^{-1} + 2z^{-2}) = X(z)$$

And so

$$H(z) = \frac{Y(z)}{X(z)} = \frac{1}{1 - 3z^{-1} + 2z^{-2}} = \frac{z^2}{z^2 - 3z + 2} = \frac{z^2}{(z-1)(z-2)}$$

Using partial fractions, we find

$$\frac{H(z)}{z} = \frac{z}{(z-1)(z-2)} = \frac{A}{z-1} + \frac{B}{z-2}$$

Therefore,

$$A = \frac{z}{z-2}\Big|_{z=1} = -1$$

and

$$B = \frac{z}{z-1}\Big|_{z=2} = 2$$

which leads to

$$H(z) = \frac{-z}{z-1} + \frac{2z}{z-2}$$

We can invert this to find the impulse response, which is

$$h[n] = -u[n] + 2(2^n u[n]) = u[n](2^{n+1} - 1)$$

Quiz

1. What is the region of convergence for the z-transform of $x[n] = \sin(\omega n)u[n]$?
2. Find the z-transform of $x[n] = \sin(\omega n)u[n]$.
3. Find the z-transform of $x[n] = \delta[n-m]$.
4. Find the z-transform of $x[n] = n^3 u[n]$.
5. Find the inverse z-transform of $X(z) = z/(z^2 - 1)$.
6. Find the inverse z-transform of $X(z) = z(z+1)/(z-1)^3$.
7. Find the z-transform of $\{2, 2, 2\}$.
8. Find the inverse z-transform of $1 - 2z^{-1} + 6z^{-2}$.
9. Find $y[n]$ if $x[n] = \{1, 1, 1\} = h[n]$.
10. Suppose that $h[n] = [2(1/2)^n - (1/4)^n]u[n]$. Find the *step response* of the system; that is, find $y[n] = x[n]*h[n]$ when $x[n] = u[n]$.

CHAPTER 12

Bode Plots

Introduction

The frequency response of a system can be plotted using a logarithmic scale in the following manner. Given the frequency response $H(\omega)$, we calculate

$$|H(\omega)|_{dB} = 20 \log_{10} |H(\omega)| \qquad (12.1)$$

We call this quantity the magnitude of the frequency response in decibels (dB). A decibel is a dimensionless unit based on the ratio of two quantities. The reader is probably familiar with the use of decibels in the study of sound. In that case, we can characterize how loud a sound is by comparing the intensity I of a given sound wave to the threshold for human hearing, which we denote by I_0. We then compute the intensity of the sound in decibels as

$$I_{dB} = 10 \log_{10} \left| \frac{I}{I_0} \right| \qquad (12.2)$$

The multiplicative constant, 10 in this example, gives us a way to compare the relative strength differences between two quantities. That is, if the difference between I_{dB1} and I_{dB2} is 10 dB, then the intensity of I_1 is 10 times the strength of I_2. Note that since the multiplicative factor in (12.1) is 20, this means that a difference of 20 dB indicates that one signal has a magnitude 20 times as large as the other.

The log function is a useful measure of signal strength for two good reasons. The first is that often quantities can vary quite a bit of strength—sometimes over many orders of magnitude. By using the logarithm we can rescale that variation down to a more manageable number. One famous example where this behavior is apparent is the Richter scale used to characterize the strength of earthquakes. The details of the Richter scale do not concern us; all that is important for our purposes is that this is a logarithmic quantity. This means that each increment on the Richter scale describes an order of magnitude increase in strength. An earthquake that is a 7 on the Richter scale is 10 times as strong as an earthquake that is a 6. In our case, using logarithms allows us to scale down a wide range of frequencies into a small scale that can be visualized and plotted more easily. This works in a similar way to the Richter scale. In our case, when the magnitude of the frequency ω increases by a factor of 10, then $\log \omega$ increases by 1.

The second reason why using logarithms is useful is that

$$\log (AB) = \log A + \log B \tag{12.3}$$

By turning multiplication into addition, the mathematics of a problem is simplified. In engineering this can be useful when calculating the overall gain of a composite system, which could be an amplifier or a filter. By using logarithms, we can simply add together the gain at each stage (in decibels) to arrive at the overall gain of the system. We call each 10-to-1 change in frequency a *decade*.

Bode Plot Basics

A Bode plot is a log-linear plot. The axes are defined in the following way:

- The horizontal axis is the logarithm of frequency ($\log_{10} \omega$).
- The vertical axis is the frequency response in decibels.

In signal analysis we will plot two quantities:

- The magnitude of the frequency response in decibels [Eq. (12.1)].
- $\theta_H(\omega)$.

In our examples, we will focus on plotting the straight line approximations to $|H(\omega)|_{dB}$ as compared to the Bode plots of the actual function, which you can easily plot using a computational math package. Our goal here is to gain a qualitative understanding of how to generate Bode plots and what they mean.

Bode Plot Examples

The key to sketching a Bode plot is to follow these steps:

- Look at very low frequency behavior (consider $\omega \to 0$)
- Look at very high frequency behavior (consider $\omega \to \infty$)
- Find the intersection with the 0-dB axis, known as the *corner frequency*

We begin with the simplest case, generating Bode plots for first-order systems.

EXAMPLE 12-1
Sketch the Bode plot for $H(\omega) = 1 + j\omega/20$.

SOLUTION 12-1
First we consider the low frequency behavior of the system. That is, we consider the magnitude of the frequency response $H(\omega)$ when $\omega \ll 20$. We have

$$|H(\omega)|_{dB} = 20\log_{10}\left|1 + j\frac{\omega}{20}\right| \to 20\log_{10}|1| \to 0 \qquad \text{as } \omega \to 0$$

Next we consider the high frequency behavior. To do this, recall that for a complex number z, the modulus is

$$|z|^2 = z\bar{z}$$

In our case, we have

$$z = 1 + j\frac{\omega}{20} \Rightarrow \bar{z} = 1 - j\frac{\omega}{20}$$

And so we get

$$|z|^2 = \left(1 + j\frac{\omega}{20}\right)\left(1 - j\frac{\omega}{20}\right)$$

$$= 1 + j\frac{\omega}{20} - j\frac{\omega}{20} + \left(j\frac{\omega}{20}\right)\left(-j\frac{\omega}{20}\right) = 1 + \frac{\omega^2}{400}$$

Now, if we consider $\omega \gg 20$, then

$$\frac{\omega^2}{400} \gg 1$$

And so we can approximate the magnitude by

$$|z|^2 = 1 + \frac{\omega^2}{400} \approx \frac{\omega^2}{400}$$

Taking the square root, $|H(\omega)| \approx \omega/20$. Therefore the large frequency expression for the magnitude expressed in decibels is

$$|H(\omega)|_{\text{dB}} \to 20 \log_{10}\left(\frac{\omega}{20}\right) \quad \text{as } \omega \to \infty$$

This is a straight line. If we set $\omega = 20$, then we have $\log_{10}(1) = 0$; so this tells us that this line intersects the 0-dB axis at $\omega = 20$, which is the corner frequency, which we'll axis at $\omega = 20$, which is the corner frequency, which we'll denote by ω_c. Combining the low frequency and high frequency behavior that we have found, we have

$$|H(\omega)|_{\text{dB}} = \begin{cases} 0 & \text{for } 0 < \omega < 20 \\ 20 \log_{10}(\omega/20) & \text{for } \omega \geq 20 \end{cases}$$

The Bode plot is just a plot of this piecewise function, shown in Fig. 12-1. Next we plot $\theta_H(\omega)$. We have

$$\theta_H(\omega) = \tan^{-1}\frac{\omega}{20}$$

The asymptotic behavior is given by

$$\theta_H(\omega) = \tan^{-1}\frac{\omega}{20} \to 0 \quad \text{as } \omega \to 0$$

$$\theta_H(\omega) = \tan^{-1}\frac{\omega}{20} \to \frac{\pi}{2} \quad \text{as } \omega \to \infty$$

A plot of $\theta_H(\omega) = \tan^{-1}(\omega/20)$ is shown in Fig. 12-2. Notice that at large frequency, $\theta_H(\omega) = \tan^{-1}(\omega/20)$ does level off at $\pi/2$.

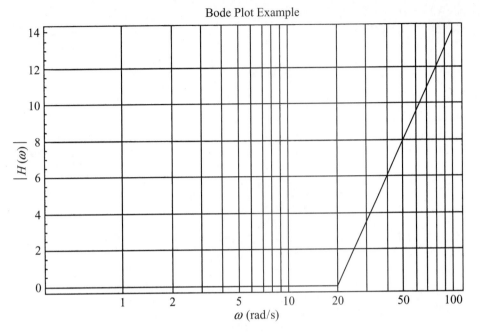

Fig. 12-1. Bode plot for Example 12-1.

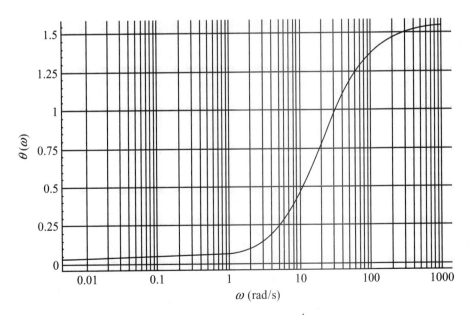

Fig. 12-2. A plot of $\theta_H(\omega) = \tan^{-1}(\omega/20)$.

EXAMPLE 12-2
Sketch the Bode plot for

$$H(\omega) = \frac{1}{1 + j\omega/10}.$$

SOLUTION 12-2
We follow the same procedure as before. We don't have to worry about the function being in the denominator, because when we take the log we can apply

$$\log \frac{1}{A} = -\log A$$

And so for $|H(\omega)|_{dB} = 20 \log_{10} |H(\omega)|$ we have

$$|H(\omega)|_{dB} = 20 \log_{10} |H(\omega)| = 20 \log_{10} \left| \frac{1}{1 + j\omega/10} \right|$$

$$= -20 \log_{10} |1 + j\tfrac{\omega}{10}|$$

Now we can proceed using the same method we applied in the last example. We can see that as $\omega \to 0$, $|H(\omega)|_{dB} \to 0$. Therefore, the low frequency behavior of this system, as defined for frequencies below the cutoff frequency, will be that the system is at a constant 0 dB.

For high frequencies, we find that for $\omega \gg 10$,

$$|H(\omega)|_{dB} = 20 \log_{10} |H(\omega)|$$

$$= 20 \log_{10} \left| \frac{1}{1 + j\omega/10} \right| \to -20 \log_{10} \left| \frac{\omega}{10} \right| \quad \text{as } \omega \to \infty$$

Once again this is a straight line, but in this case we have a minus sign out front, giving a negative slope. The corner frequency is given by $\omega = 10$, and so we have

$$|H(\omega_c)|_{dB} = 20 \log_{10} |H(\omega_c)| = 20 \log_{10} \left| \frac{1}{1 + j(10/10)} \right|$$

$$= 20 \log_{10} \left| \frac{1}{1 + j(1)} \right| = -20 \log_{10} |1 + j|$$

$$= -20 \log_{10} \sqrt{2} = -3 \text{ dB}$$

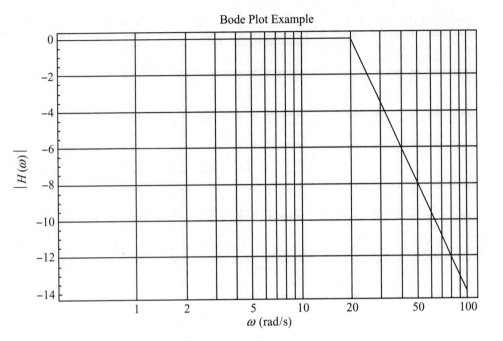

Fig. 12-3. The linearly decreasing case in Example 12-2.

Putting these results together, we see that the system response decreases with increasing frequency. This is shown in Fig. 12-3.

To characterize $\theta_H(\omega)$, we again pick up a minus sign since the function of frequency in this case is in the denominator. So we have

$$\theta_H(\omega) = -\tan^{-1}\frac{\omega}{10}$$

With asymptotic behavior given by

$$\theta_H(\omega) = -\tan^{-1}\frac{\omega}{10} \to 0 \qquad \text{as } \omega \to 0$$

$$\theta_H(\omega) = -\tan^{-1}\frac{\omega}{10} \to -\frac{\pi}{2} \qquad \text{as } \omega \to \infty$$

A plot of this is shown in Fig. 12-4.

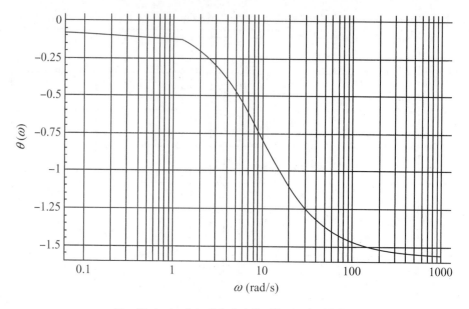

Fig. 12-4. A plot of $\theta_H(\omega)$ for Example 12-2.

EXAMPLE 12-3
For our final example, sketch the Bode plot for

$$H(\omega) = 300 \frac{5 + j\omega}{-\omega^2 + j11\omega + 10}$$

SOLUTION 12-3
We begin by rewriting the transfer function in a more convenient form:

$$H(\omega) = 300 \frac{5 + j\omega}{-\omega^2 + j11\omega + 10} = 300 \frac{5 + j\omega}{(1 + j\omega)(10 + j\omega)}$$

Now let's factor out the 10 in the denominator and the 5 in the numerator:

$$H(\omega) = 300 \frac{5 + j\omega}{(1 + j\omega)(10 + j\omega)} = 300 \frac{5 + j\omega}{(1 + j\omega)(10)(1 + j\omega/10)}$$

$$= 150 \frac{1 + j\omega/5}{(1 + j\omega)(1 + j\omega/10)}$$

When we compute the logarithm, we can use $\log(A/BC) = \log A - \log B - \log C$. And so we have

$$|H(\omega)|_{dB} = 20 \log_{10} \left| 150 \frac{1 + j\omega/5}{(1 + j\omega)(1 + j\omega/10)} \right|$$

$$= 20 \log_{10} |150| + 20 \log_{10} |1 + j\omega/5|$$

$$- 20 \log_{10} |1 + j\omega| - 20 \log_{10} |1 + j\omega/10|$$

The first term is just a constant. For the other three terms, notice that there are three corner frequencies. We consider each in turn. The corner frequency for $20 \log_{10} |1 + j\omega/5|$ is given by $\omega_{c1} = 1$ and we have

$$20 \log_{10} |150| + 20 \log_{10} |1 + j/5| - 20 \log_{10} |1 + j| - 20 \log_{10} |1 + j\omega/10|$$

$$= 20 \log_{10} |150| + 20 \log_{10} \sqrt{26/25} - 20 \log_{10} \sqrt{2} - 20 \log_{10} \sqrt{101/100}$$

$$\approx 40.6 \text{ dB}$$

The next corner frequency is given by $\omega_{c2} = 5$ where we find that

$$20 \log_{10} |150| + 20 \log_{10} |1 + j| - 20 \log_{10} |1 + j5| - 20 \log_{10} |1 + j\omega/2|$$

$$= 20 \log_{10} |150| + 20 \log_{10} \sqrt{2} - 20 \log_{10} \sqrt{26} - 20 \log_{10} \sqrt{5/4} \approx 30.4 \text{ dB}$$

Finally, the last corner frequency is at $\omega_{c3} = 10$. In this case we find

$$20 \log_{10} |150| + 20 \log_{10} |1 + j2| - 20 \log_{10} |1 + j10| - 20 \log_{10} |1 + j|$$

$$= 20 \log_{10} |150| + 20 \log_{10} \sqrt{5} - 20 \log_{10} \sqrt{101} - 20 \log_{10} \sqrt{2} \approx 27.5 \text{ dB}$$

Returning to the original expression, we had

$$|H(\omega)|_{dB} = 20 \log_{10} |150| + 20 \log_{10} |1 + j\omega/5|$$

$$- 20 \log_{10} |1 + j\omega| - 20 \log_{10} |1 + j\omega/10|$$

The first term is $20 \log_{10} |150| \approx 44$ dB which adds a constant or piston term to the plot. To generate the Bode plot, we add each term at the appropriate corner frequency. Since the first corner frequency occurs at $\omega_{c1} = 1$, up to that point we have the constant term

$$|H(\omega)|_{dB} = 20 \log_{10} |150| \quad 0 < \omega \le 1$$

Next, between $\omega_{c1} = 1$ and $\omega_{c2} = 5$, we add the second term whose corner frequency is $\omega_{c2} = 5$, giving

$$|H(\omega)|_{dB} = 20 \log_{10} |150| - 20 \log_{10} |1 + j\omega| \quad 1 < \omega \leq 5$$

The next corner frequency occurs at $\omega_{c3} = 10$. Up until this point we add in the next term

$$|H(\omega)|_{dB} = 20 \log_{10} |150| + 20 \log_{10} |1 + j\omega/5|$$
$$- 20 \log_{10} |1 + j\omega| \quad 5 < \omega \leq 10$$

The last part of the plot is for $\omega > 10$, where we add in the last term to obtain

$$|H(\omega)|_{dB} = 20 \log_{10} |150| + 20 \log_{10} |1 + j\omega/5|$$
$$-20 \log_{10} |1 + j\omega| - 20 \log_{10} |1 + j\omega/10|$$

The plot is shown below in Fig. 12-5.

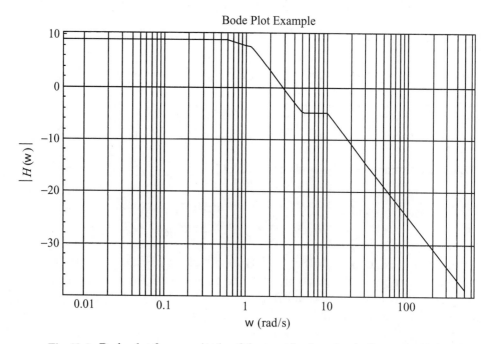

Fig. 12-5. Bode plot for magnitude of the transfer function in Example 12-3.

Quiz

1. Plot $|H(\omega)|_{dB}$ for $H(\omega) = 1 + j\omega/30$.
2. What are the corner frequencies for

$$H(\omega) = 1000 \times \frac{1 + j\omega}{1000 + 110j\omega - \omega^2}.$$

3. Construct a Bode plot for the transfer function of Problem 2.

FINAL EXAM

1. An energy signal is best defined by
    (a) $0 < E < \infty$ and $0 < P < \infty$
    (b) $0 \leq E < \infty$
    (c) $0 < E < \infty$ and $P = 0$
    (d) $0 < P < \infty$ and $E = 0$

2. The total energy content of $t^3 \sin \pi t$ defined for $0 \leq t \leq \pi$ is about
    (a) 137
    (b) 147
    (c) 141
    (d) 150

3. The total energy content of $x(t) = t^2 e^{-2t}$, $0 \leq t$ is about
    (a) 0.023
    (b) 0.000
    (c) 0.230
    (d) $E \to \infty$

4. Let $x(t) = 2[u(t + \alpha) - u(t - \alpha)]$. What is the total power?
    (a) 4α
    (b) 2α
    (c) $P \to \infty$
    (d) 4

5. The best way to describe a time-invariant system is to say that
    (a) the response $y(t)$ is constant
    (b) if the response to $x(t)$ is $y(t)$, then the response to $x(t - a)$ is $y(t - a)$
    (c) if the response to $x(t)$ is $y(t)$, then the response to $x(t - a)$ is $y(t + a)$
    (d) if the response to $x(t)$ is $y(t)$, then the response to $x(t - a)$ is $y(at)$

6. The fundamental period is defined as
    (a) the smallest value T_0 such that $x(t) = x(t + T_0)$
    (b) the time at which a system is time-invariant
    (c) the time at which the output of a system exhibits resonance with the input signal

7. Consider a periodic signal. If the energy content over one period is finite, then the power content is defined by
    (a) $P = 0$
    (b) $P = E_0 / T_0$
    (c) $P = E_0 T_0$
    (d) Power is undefined for periodic signals

8. Euler's formula allows us to write
    (a) $\cos \omega t = (e^{j\omega t} + e^{-j\omega t})/2j, \quad \sin \omega t = (e^{j\omega t} - e^{-j\omega t})/2j$
    (b) $\cos \omega t = (e^{j\omega t} - e^{-j\omega t})/2, \quad \sin \omega t = (e^{j\omega t} - e^{-j\omega t})/2j$
    (c) $\cos \omega t = (e^{j\omega t} + e^{-j\omega t})/2, \quad \sin \omega t = (e^{j\omega t} - e^{-j\omega t})/2j$
    (d) $\cos \omega t = (e^{j\omega t} - e^{-j\omega t})/2j, \quad \sin \omega t = (e^{j\omega t} + e^{-j\omega t})/2$

9. The energy content of $x(t) = e^{-8t}u(t)$ is
    (a) 1/8
    (b) 1/4
    (c) 1/16
    (d) 0

10. The even part of $x(t) = e^{-2t}$ is
    (a) $\sinh t$
    (b) e^{2t}
    (c) $\cos 2t$
    (d) $\cosh 2t$

11. The system $y(t) = \int_{-\infty}^{t} x(\tau) \, d\tau$ can be described as
    (a) memoryless
    (b) has memory
    (c) not enough information has been provided to determine the memory of this system

12. The best description of a causal system is
    (a) the system output $y(t)$ depends only on the input at present or earlier times
    (b) the system output $y(t)$ depends only on the input at the present time
    (c) the system output $y(t)$ depends only on the input at earlier times
    (d) the system output $y(t)$ depends on future values of the input

13. The system $y(t) = 4x^2(t) - 2x(t)$ is
(a) linear
(b) nonlinear

14. One way to describe a system as stable is
(a) the output remains constant with time
(b) the input is always bounded
(c) if an input signal is bounded, then the output signal is also bounded
(d) the system is memoryless

15. The system $y(t) = x(t)\cosh t$ is
(a) memoryless and causal
(b) only memoryless
(c) BIBO stable

16. The system described by $y(t) = \int_{-\infty}^{t/2} x^2(s)\,ds$ is
(a) memoryless and causal
(b) not memoryless

17. The sampling property of the Dirac delta function is
(a) $\int_{-\infty}^{\infty} \delta(t)\,dt = 1$
(b) $\int_{-\infty}^{a} \delta(t)\,dt = a$
(c) $\int_{-\infty}^{\infty} \phi(t)\delta(t-a)\,dt = \phi(a)$

18. If you evaluate $\int_{0}^{\infty} e^{-t}\delta(t-7)\,dt$ you get
(a) 0
(b) 1
(c) e^{-7}
(d) $e^{7}\delta(t-7)$

19. Which of the following is correct?
(a) $\int_{-\infty}^{\infty} u(t-2)x(t)\,dt = \int_{2}^{\infty} x(t)\,dt$
(b) $\int_{-\infty}^{\infty} u(t-2)x(t)\,dt = \int_{-\infty}^{\infty} \delta(t-2)x(t)\,dt = x(2)$
(c) $\int_{-\infty}^{\infty} u(t-2)x(t)\,dt = \int_{0}^{\infty} x(t)\,dt$
(d) $\int_{-\infty}^{\infty} u(t-2)x(t)\,dt = \int_{-\infty}^{2} x(t)\,dt$

20. The unit step function is related to the unit impulse function via which of the following relationships:
(a) $u(t) = \int_{0}^{t} \delta(\tau)\,d\tau$
(b) $u(t) = \int_{-\infty}^{t} \delta(\tau)\,d\tau$

(c) $u(t) = \int_t^\infty \delta(\tau)\,d\tau$

(d) $u(t) = \sum_{n=0}^\infty \delta(t-n)$

21. The step response of a system can be written as
   (a) $s(t) = u(t)*h(t) = \int_{-\infty}^\infty u(\tau)h(t-\tau)\,d\tau$
   (b) $s(t) = \int_{-\infty}^\infty \delta(t)h(t-\tau)\,d\tau$
   (c) $s(t) = u(t)x(t)$

22. If the impulse response of a linear time-invariant system satisfies $\int_{-\infty}^\infty |h(\tau)|\,d\tau < \infty$ we say that
   (a) the system is noncausal
   (b) the sytem is causal
   (c) the system is stable
   (d) the system has memory

23. A linear time-invariant system that is causal is one for which
   (a) $h(t) = 0$ when $t < 0$
   (b) $\int_0^\infty h(t)\,dt = 0$
   (c) $h(t) \le 0$ when $t < 0$

24. The linear time-invariant system with $h(t) = 4e^{-2t}u(t)$ is
   (a) not stable
   (b) stable, causal, but has memory
   (c) stable, causal, and memoryless
   (d) stable, but not causal

25. The convolution operation is
   (a) anticommutative, meaning $f*g = -g*f$
   (b) commutative $f*g = g*f$ provided that g is real
   (c) anticommutative if f is real
   (d) always commutative, so $f*g = g*f$

26. In a discrete-time signal, the most general way to write the energy is
   (a) $E = \sum_{-\infty}^\infty |x[n]|^2$
   (b) $E = \sum_0^\infty |x[n]|^2$
   (c) $E = \frac{1}{N}\sum_{-\infty}^\infty |x[n]|^2$
   (d) $E = \sum_{-\infty}^\infty |x[n]|$

27. A discrete-time signal is an energy signal if
   (a) discrete-time signals are only power signals
   (b) $0 < P < \infty$ and E is finite

(c) $0 < E < \infty$ and $P = 0$

(d) discrete-time signals are neither energy or power signals

28. If $\{x_n\} = \sum_{n=0}^{\infty} ar^n$ and $|r| < 1$, the energy content of the signal is
    (a) $E = a/(1 - r)$
    (b) $E = a/(1 - r^2)$
    (c) $E \rightarrow \infty$
    (d) $E = a$

29. The average power of a discrete-time signal which is zero for $n < 0$ is
    (a) $P = \lim_{N \rightarrow \infty} \frac{1}{2N+1} \sum_{n=0}^{N} |x[n]|^2$
    (b) $P = \lim_{N \rightarrow \infty} \frac{1}{N} \sum_{n=0}^{N} |x[n]|^2$
    (c) $P = \lim_{N \rightarrow \infty} \frac{1}{2N+1} \sum_{n=0}^{N} \frac{|x[n]|^2}{n^2}$

30. A discrete-time sequence can be written as
    (a) $x[n] = \sum_{k=-\infty}^{\infty} x[k]u[n - k]$
    (b) $x[n] = \sum_{k=-\infty}^{\infty} x[k]h[n - k]$
    (c) $x[n] = \sum_{k=-\infty}^{n} x[k]\delta[k]$
    (d) $x[n] = \sum_{k=-\infty}^{\infty} x[k]\delta[n - k]$

31. The relationship between fundamental frequency ω_0 and fundamental period N_0 is
    (a) $\omega_0 = 2\pi/N_0$
    (b) $\omega_0 = N_0/2$
    (c) $\omega_0 = N_0/2\pi$
    (d) $N_0 = 2\pi/\omega_0$

32. A signal $x(t)$ is said to be even if
    (a) $x(t) = -x(-t)$
    (b) while functions can said to be even or odd, this definition does not apply to signals since they are strictly zero if $t < 0$
    (c) $x(t) + x(-t) = 2$
    (d) $x(t) = x(-t)$

33. The system described by $y[n] = 3x[n - 1]$ is
    (a) causal
    (b) not causal

34. The system described by $y[n] = 3x[n + 1]$ is
    (a) causal
    (b) not causal

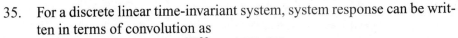
35. For a discrete linear time-invariant system, system response can be written in terms of convolution as
    (a) $y[n] = h[n]^*x[n] = \sum_{k=-\infty}^{\infty} h[k]x[k]$
    (b) $y[n] = h[n]^*x[n] = \sum_{k=-\infty}^{\infty} h[n+k]x[n-k]$
    (c) $y[n] = h[n]^*x[n] = \sum_{k=-\infty}^{\infty} h[k]x[n+k]$
    (d) $y[n] = h[n]^*x[n] = \sum_{k=-\infty}^{\infty} h[k]x[n-k]$

36. A memoryless discrete linear time-invariant system can be written as
    (a) $y[n] = Kx[n]$
    (b) $y[n] = Kx[-n]$
    (c) $y[n] = \delta[m-n]x[n]$
    (d) $y[n] = u[n]x[n]$

37. A discrete time linear time-invariant system is Bounded Input, Bounded Output (BIBO) stable if
    (a) $\sum_{k=-\infty}^{\infty} |y[k]| < \infty$
    (b) $\sum_{k=-\infty}^{\infty} |h[k]| < \infty$
    (c) $\sum_{k=-\infty}^{\infty} |y[k]| \leq \sum_{k=-\infty}^{\infty} |h[k]|$
    (d) $\sum_{k=-\infty}^{\infty} |y[k]| \leq \sum_{k=-\infty}^{\infty} |x[k]|$

38. A discrete linear time-invariant system with $h[n] = nu[n]$ is
    (a) memoryless and causal
    (b) memoryless but not causal
    (c) not memoryless, but causal
    (d) stable

39. The Fourier series expansion of a periodic signal can be written as
    (a) $x(t) = a_0 + 2\sum_{n=1}^{\infty} [a_n \cos(2\pi nt/T_0) + b_n \sin(2\pi nt/T_0)]$
    (b) $x(t) = a_0 + 2\sum_{n=1}^{\infty} [a_n \cos(2\pi nt/T_0) - b_n \sin(2\pi nt/T_0)]$
    (c) $x(t) = a_0 - 2\sum_{n=1}^{\infty} [a_n \cos(2\pi nt/T_0) + b_n \sin(2\pi nt/T_0)]$
    (d) $x(t) = a_0 + 2\sum_{n=1}^{\infty} [a_n \cos(2\pi nt/T_0) + b_n \sin(2\pi nt/T_0)]$

40. The Fourier transform of $x(t) = e^{-t} \cos t$ is given by
    (a) $\pi\delta(j-1-2\pi f) + \pi\delta(j+1-2\pi f)$
    (b) $\pi\delta(2\pi f)$
    (c) $e^{-1}\pi[\delta(j-1-2\pi f) + \delta(j+1-2\pi f)]$

41. The Fourier transform of $u(t)e^{-t}$ is
    (a) $j/(j+2\pi f)$
    (b) $j/(j-f)$
    (c) $j/(j-2\pi f)$
    (d) $2\pi/(j-2\pi f)$

42. The Fourier transform of $x(t) = 1/t$ is
    (a) $-j\pi u(f)$
    (b) $-j\pi\, sgn(f)$
    (c) $-\pi\delta(f)$

43. The mean value of a periodic signal can be found using the Fourier series expansion and
    (a) calculating the expansion coefficient a_0
    (b) averaging over the a_n
    (c) averaging over the b_n

44. The Fourier transform of $x(t) = t$ is given by
    (a) $\frac{j}{2\pi}u'(f)$
    (b) $\frac{j}{2\pi}\delta(f)$
    (c) $\frac{j}{2\pi}\delta'(f)$
    (d) it cannot be calculated, the integral is divergent

45. The Dirichlet conditions for the existence of a Fourier series require that
    (a) the signal $x(t)$ has no discontinuities over the fundamental period
    (b) The signal $x(t)$ has a finite number of discontinuities, minima, and maxima over the fundamental period
    (c) the discontinuities of $x(t)$ are differentiable
    (d) the signal $x(t)$ has no global minima or maxima over the fundamental period

46. The complex exponential Fourier series representation is given by
    (a) $x(t) = \sum_{-\infty}^{\infty} c_n \exp(j\frac{2\pi nt}{T_0})$
    (b) $x(t) = a_0 + \sum_{-\infty}^{\infty} c_n \exp(j\frac{2\pi nt}{T_0})$
    (c) $x(t) = a_0 - \sum_{-\infty}^{\infty} c_n \exp(j\frac{2\pi nt}{T_0})$
    (d) $x(t) = \sum_{-\infty}^{\infty} c_n \exp(-j\frac{2\pi nt}{T_0})$

47. Using Parseval's theorem, the power in a periodic signal can be written as
    (a) $P = \frac{1}{T_0}\int_{-T_0/2}^{T_0/2} |x(t)|\, dt = \sum_{n=-\infty}^{\infty} |c_n|$
    (b) $P = \frac{1}{T_0}\int_{-T_0/2}^{T_0/2} |x(t)|^2\, dt = 0$
    (c) $P = \frac{1}{T_0}\int_{-T_0/2}^{T_0/2} |x(t)|^2\, dt = \sum_{n=-\infty}^{\infty} |c_n|^2$

48. In order for a signal to be Fourier transformable, it must satisfy
    (a) $\int_{-\infty}^{\infty} |x(t)|\, dt < \infty$
    (b) $\int_{-\infty}^{\infty} |x(t)|\, dt < 0$

(c) $\int_{-\infty}^{\infty} |x(t)|^2 \, dt < 0$

(d) $\int_{-\infty}^{\infty} |x(t)|^2 \, dt < \infty$

49. The Fourier transform of a unit impulse function is a member of the Fourier transform pair

(a) $\begin{aligned} \delta(t) &\rightleftharpoons 1 \\ \delta(t-a) &\rightleftharpoons \exp(-j2\pi f) \end{aligned}$

(b) $\begin{aligned} \delta(t) &\rightleftharpoons 1 \\ \delta(t-a) &\rightleftharpoons \exp(-j2\pi f a) \end{aligned}$

(c) $\begin{aligned} \delta(t) &\rightleftharpoons 1 \\ \delta(t-a) &\rightleftharpoons \exp(j2\pi f a) \end{aligned}$

50. The Fourier transform of $u(t)e^{-2t} \cos 5t$ is

(a) $\dfrac{1+j\pi f}{29 + j8\pi f - 4\pi^2 f^2}$

(b) $\dfrac{2 + j2\pi f}{29 + j8\pi f - 4\pi^2 f^2}$

(c) $\dfrac{2 + j2\pi f}{25 + j8\pi f - 4\pi^2 f^2}$

(d) $\dfrac{5 + j5\pi f}{2 + j8\pi f - 4\pi^2 f^2}$

51. The Fourier transform of a Gaussian pulse is
(a) a Gaussian pulse in frequency
(b) a decaying exponential
(c) a Gaussian pulse multiplied by the unit step function
(d) zero because it is symmetric

52. The time-shifting property of the Fourier transform can be described as
(a) $FT[x(t-a)] = e^{-jaf} X(f)$
(b) $FT[x(t-a)] = e^{-2\pi af} X(f)$
(c) $FT[x(t-a)] = e^{-j2\pi af} X(f)$
(d) $FT[x(t-a)] = e^{j2\pi af} X(f)$

53. The time-scaling property of the Fourier transform can be described as
(a) $x(at) \rightleftharpoons \frac{1}{|a|} X(af)$
(b) $x(at) \rightleftharpoons X(\frac{f}{a})$
(c) $x(\frac{t}{a}) \rightleftharpoons \frac{1}{|a|} X(af)$
(d) $x(at) \rightleftharpoons \frac{1}{|a|} X(\frac{f}{a})$

54. Convolution in the time domain corresponds to
   (a) multiplication in the frequency domain
   (b) time-shifted convolution in the frequency domain
   (c) scaled convolution in the frequency domain
   (d) multiplication in the frequency domain, with the second signal shifted in frequency

55. The amplitude spectrum of a signal $x(t)$ is
   (a) $X(f) = |X(f)|e^{j\phi}$
   (b) $|X(f)|$, the magnitude of the Fourier transform
   (c) $\phi = \arg(X(f))$

56. Parseval's theorem states that
   (a) $\int_{-\infty}^{\infty} |x(t)|^2\, dt = \frac{1}{2\pi} \int_{-\infty}^{\infty} |X(\omega)|^2\, d\omega$
   (b) $\int_{-\infty}^{\infty} |x(t)|\, dt = \frac{1}{2\pi} \int_{-\infty}^{\infty} |X(\omega)|\, d\omega$
   (c) $\frac{1}{2\pi} \int_{-\infty}^{\infty} |X(\omega)|^2\, d\omega < \infty$
   (d) $\int_{-\infty}^{\infty} |x(t)|^2\, dt < \infty$

57. The response of a linear time-invariant system can be written as
   (a) $Y(f) = H(f)^* X(f)$
   (b) $Y(f) = X(f)H(-f)$
   (c) $Y(f) = X(f)H(f)$
   (d) $Y(f) = X(f_a)\delta(f - f_a)$

58. The relationship between the phase of an input signal and the phase of an output signal in the frequency domain for a linear time-invariant system can be written as
   (a) $\theta_Y(f) = \theta_X(f) + \theta_H(f)$
   (b) $\theta_Y(f) = \theta_X(f) - \theta_H(f)$
   (c) $\theta_Y(f) = \theta_X(f)\theta_H(f)$
   (d) $\theta_Y(f) = \theta_X(f) \otimes \theta_H(f)$

59. If $h(t) = 5e^{-t}u(t)$ then the frequency response is
   (a) $5\delta(2\pi f)$
   (b) $H(f) = 5/(5 + j2\pi f)$
   (c) $H(f) = 1/(5 + j2\pi f)$
   (d) $H(f) = 5/(1 + j2\pi f)$

60. For a system with distortionless transmission, the frequency response has
   (a) amplitude that varies linearly with the gain constant
   (b) zero gain constant

(c) gain constant equal to unity

(d) constant amplitude for all frequencies that is equal to the gain constant

61. An ideal low pass filter has

(a) a frequency response that satisfies $|H(f)| = \begin{cases} 1 & |f| < f_c \\ 0 & |f| > f_c \end{cases}$

(b) a frequency response that satisfies $|H(f)| = \begin{cases} -1 & |f| < f_c \\ 0 & |f| > f_c \end{cases}$

(c) a response that satisfies $|Y(f)| = \begin{cases} 1 & |f| < f_c \\ 0 & |f| > f_c \end{cases}$

62. The *bandwidth* of a low pass filter is defined by

(a) the maximum frequency of the input signal

(b) the cutoff frequency

(c) the lowest frequency in the input signal

(d) the central frequency of the impulse function

63. For an ideal low pass filter, the phase spectrum

(a) is linearly decreasing between $\pm f_c$

(b) is linearly increasing between $\pm f_c$

(c) vanishes between $\pm f_c$

(d) is constant between $\pm f_c$

64. Ideal filters

(a) are always causal

(b) are always memoryless and causal

(c) are not causal

(d) have vanishing delta response

65. The midband frequency for a band pass filter can be written as

(a) $\omega_0 = \frac{1}{2}(\omega_1 - \omega_2)$

(b) $\omega_0 = \omega_1/\omega_2$

(c) $\omega_0 = 0$

(d) $\omega_0 = \frac{1}{2}(\omega_1 + \omega_2)$

66. Energy spectral density

(a) is energy per unit bandwidth

(b) vanishes for linear systems

(c) can be calculated using convolution in the frequency domain

(d) is constant in linear time-invariant systems

67. The cross-correlation function can be written as

(a) $R_{12}(\tau) = \lim\limits_{T \to \infty} \int_{-T}^{T} x_1(t)x_2(t)\,dt$

(b) $R_{12}(\tau) = \lim\limits_{T \to \infty} \int_{-T}^{T} x_1(t)x_2(t - \tau)\,dt$

(c) $R_{12}(\tau) = \lim\limits_{T \to \infty} \int_{-T}^{T} x_1(t + \tau)x_2(t)\,dt$

(d) $R_{12}(\tau) = \lim\limits_{T \to 0} \int_{-T}^{T} x_1(t)x_2(t - \tau)\,dt$

68. Given a periodic signal, the autocorrelation function
(a) has the same period as the signal
(b) has half the period as the signal
(c) has twice the period as the signal
(d) is nonperiodic

69. Given a power signal $x(t)$, the power spectral density is
(a) negative
(b) nonnegative, but may be complex
(c) nonnegative and real valued

70. The power spectral density of a real-valued power signal satisfies
(a) $S(f) = jS(-f)$
(b) $S(f) = S(-f)$
(c) $S(f) = -S(-f)$
(d) $S(f) = -jS(-f)$

71. The discrete Fourier transform of $\{2, -1, 0, 0, 2, 1\}$ is given by
(a) $\{4, 1 + j3.46, 0, 1, 1, 1 - j3.46\}$
(b) $\{4, 1 + j3.46, 1, 4, 1, 1 - j3.46\}$
(c) $\{4, 1 + j3.46, 1, 4, 1, 1 + j3.46\}$
(d) $\{4, 1 + j3.46, 0, 4, 1, 1 - j3.46\}$

72. The discrete Fourier transform of $\{2, 1, 0\}$ is given by
(a) $\{3, 1.5 - j8.67, 1.5 + j8.67\}$
(b) $\{3, 1.5 + j8.67, 1.5 - j8.67\}$
(c) $\{3, 0, 1.5 + j8.67\}$
(d) $\{3, 0, 0\}$

73. If a discrete signal is periodic then the coefficients in its Fourier series expansion satisfy
(a) they vanish outside the fundamental period
(b) they are constant outside the fundamental period
(c) $c_k = c_{k+N_0}$
(d) $c_k = -c_{k+N_0}$

74. The inverse Fourier transform of a *discrete* signal is given by
(a) $x[n] = \frac{1}{2\pi} \sum_{2\pi} X(\Omega)e^{j\Omega n}$
(b) $x[n] = \frac{1}{2\pi} \int_{2\pi} X(\Omega)e^{j\Omega n}\, d\Omega$
(c) $x[n] = \frac{1}{2\pi} \int_{-\infty}^{\infty} X(\Omega)e^{j\Omega n}\, d\Omega$

75. The discrete Fourier transform can be written as
(a) $X[k] = \sum_{n=0}^{N-1} x[n]e^{-j\frac{2\pi}{N}kn}$
(b) $X[k] = \sum_{n=0}^{N-1} x[n]e^{j\frac{2\pi}{N}kn}$
(c) $X[k] = a_0 + \sum_{n=0}^{N-1} x[n]e^{-j\frac{2\pi}{N}kn}$
(d) $X[k] = \sum_{n=0}^{N-1} x^2[n]e^{-j\frac{2\pi}{N}kn}$

76. The inverse discrete Fourier transform of $\{2,\ j,\ -j\}$ is given by
(a) $\{0.67,\ 0.09,\ 1.24\}$
(b) $\{0.87,\ 0.09,\ 1.24\}$
(c) $\{0.67,\ 0,\ 1.24\}$
(d) $\{0,\ 0.09,\ 1.24\}$

77. The period of a 550-Hz signal is
(a) 5.5 ms
(b) 4 ms
(c) 1.82 ms
(d) 1.82 s

78. If a signal is bandlimited to $[-\omega/2,\ \omega/2]$ and the sampling rate is T_s then the Nyquist frequency is
(a) $1/2T_s$
(b) $1/T_s$
(c) $2/T_s$

79. The phenomenon of leakage is best described by
(a) a decaying phase spectrum
(b) the spreading or leakage of frequencies in the discrete Fourier transform
(c) a spreading amplitude spectrum
(d) is associated with sinusoidal signals only

80. The modulation index for amplitude modulation is given by
(a) $\mu = -|\max\{m(t)\}|/A_c$
(b) $\mu = |\max\{m(t)\}|$
(c) $\alpha = -|\max\{m(t)\}|$
(d) $\mu = |\max\{m(t)\}|/A_c$

81. An amplitude-modulated wave can be demodulated if
    (a) $\mu \leq 1$
    (b) $\mu < 1$
    (c) $\mu = 0$
    (d) $\mu \geq 1$

82. Percent modulation is defined as
    (a) $|\max\{m(t)\}|(100)$
    (b) $\frac{k_a}{\mu}|\max\{m(t)\}|(100)$
    (c) $\frac{1}{\mu}(100)$
    (d) $k_a|\max\{m(t)\}|(100)$

83. If $x_c(t) = 3\cos(20\pi t)$ then the carrier frequency is
    (a) 20 Hz
    (b) 10 Hz
    (c) 2π Hz
    (d) 20π Hz

84. In an amplitude-modulated wave, carrier power is defined as
    (a) $P_c = \frac{1}{2}A_c^2$
    (b) $P_c = \frac{1}{2}A_c$
    (c) $P_c = A_c^2$
    (d) $P_c = \frac{1}{2}A_c^2 - \mu$

85. The total power in an amplitude-modulated wave is given by
    (a) $P_t = P_c + P_s = \frac{1}{2}(1 - \frac{1}{2}\mu^2)A_c^2$
    (b) $(1 + \mu^2)A_c^2$
    (c) $P_t = (\frac{1}{2}\mu^2)A_c^2$
    (d) $\frac{1}{2}(1 + \frac{1}{2}\mu^2)A_c^2$

86. The efficiency of an amplitude-modulated wave is
    (a) $\eta = -P_s/P_t$
    (b) $\eta = P_s/P_t$
    (c) $\eta = (P_s/P_t) \times 100\%$
    (d) $\eta = (P_t/P_s) \times 100\%$

87. In the frequency domain, an ordinary or standard amplitude-modulated signal can be written as
    (a) $\pi A_c[\delta(\omega - \omega_c) + \delta(\omega + \omega_c)] + \frac{1}{2}M(\omega - \omega_c) + \frac{1}{2}M(\omega + \omega_c)$
    (b) $\pi A_c[\delta(\omega - \omega_c) + \delta(\omega + \omega_c)]M(\omega)$
    (c) $\pi A_c[\delta(\omega - \omega_c) + \delta(\omega + \omega_c)] - \frac{1}{2}M(\omega - \omega_c) + \frac{1}{2}M(\omega + \omega_c)$

88. The instantaneous frequency is
    (a) $\omega_i = d\theta/dt = \omega_c - (d\phi/dt)$
    (b) $\omega_i = d\phi\ dt$
    (c) $\omega_i = d\theta/dt = \omega_c + (d\phi/dt)$

89. Phase modulation is defined by
    (a) $\phi(t) = k_p m(t)$
    (b) $d\phi/dt = k_p m(t)$
    (c) $\phi(t) = -k_p(dm/dt)$

90. Frequency modulation is defined by
    (a) $\phi = k_f m(t)$
    (b) $d\phi/dt = k_f m(t)$
    (c) $d\phi/dt = k_f(d^2 m/dt^2)$
    (d) $d\phi/dt = k_f(d^2 m/dt^2) + (dm/dt)$

91. The carrier signal for a frequency-modulated wave can be written as
    (a) $x_{FM}(t) = A_c \cos[\omega_c t + k_f \phi - \frac{dm}{dt}]$
    (b) $x_{FM}(t) = A_c \cos[\omega_c t + k_f \int_0^t m(\tau)\,d\tau]$
    (c) $x_{FM}(t) = A_c \cos[\omega_c t + k_f \frac{dm}{dt}]$

92. The instantaneous frequency of $x(t) = 2\,\cos(3\pi t + \frac{\pi}{4})$ is
    (a) 0
    (b) $3\pi t$
    (c) 2 Hz
    (d) 1.5 Hz

93. The z-Transform of a discrete time signal is given by
    (a) $X(z) = \sum_{n=-\infty}^{\infty} x[n]z^{-n}$
    (b) $X(z) = \sum_{n=-\infty}^{\infty} x[n]z^{n}$
    (c) $X(z) = \sum_{n=-\infty}^{\infty} x[n]z$
    (d) $X(z) = \sum_{n=-\infty}^{\infty} x[n]z^{-n-1}$

94. The z-Transform of 1 is
    (a) $1/(z-1)$
    (b) $z/(z-1)$
    (c) $z/(z+1)$
    (d) $1/z$

95. The z-Transform of n is
    (a) $z/(z-1)^2$

(b) $z/(z-1)$

(c) $z/(z-1)^3$

96. The inverse z-Transform of $z/(z-2)$ is

(a) n^2

(b) $2n$

(c) 2^{n-1}

(d) 2^n

97. The Laplace transform of $x(t) = \cosh t$ is

(a) $s/(s^2-1)$

(b) $s/(s^2+1)$

(c) $s/(s^2+1)$

(d) $1/(s^2+1)$

98. The Laplace transform of $e^t \sinh t$ is given by

(a) $\dfrac{1}{s(s+2)}$

(b) $\dfrac{1}{s(s-2)}$

(c) $\dfrac{1}{s(s^2-2)}$

99. If $X(s) = s$ then $x(t)$ is

(a) $\dfrac{d}{dt}\delta(t)$

(b) $\delta(t)$

(c) $e^{-t}\delta(t)$

100. If $X(s) = 1/(s^2 + 2s - 8)$ then $x(t)$ is

(a) $\frac{1}{12}(e^{2t} - e^{4t})$

(b) $\frac{1}{6}(e^{-2t} - e^{-4t})$

(c) $\frac{1}{6}(e^{2t} + e^{-4t})$

(d) $\frac{1}{6}(e^{2t} - e^{-4t})$

QUIZ SOLUTIONS

Chapter 1

1. The signal is a power signal, with $E = \infty$ and $P = 1/2$
2. $E = 2.81$
3. $E = 16/3, \ P = 0$
4. $E_0 = \pi/2\omega^3$
5. $P = 1/4\omega^2$
6. $x_e(t) = \cos t \sin^2 t, \ x_o(t) = \sin t \cos^2 t$
7. Odd
8. $E = 0.225$
9. $T_0 = 2\pi/\omega$
10. $x[n] = x[n + N]$

Chapter 2

1. The system is memoryless, but is not time invariant since $y(t - \tau) = (t - \tau)x(t - \tau)$ but $\hat{T}\{x(t - \tau)\} = tx(t - \tau)$.
2. No, because $\hat{T}\{x_1 + x_2\} = 1/(x_1 + x_2)$, but $y_1 + y_2 = (1/x_1) + (1/x_2) = (x_1 + x_2)/x_1x_2$
3. No the system is not time-invariant
4. The system is linear. Consider

$$y_1 + y_2 = t^2\left(\frac{dx_1}{dt} + \frac{dx_2}{dt}\right) = t^2\frac{d}{dt}(x_1 + x_2) = \hat{T}\{x_1 + x_2\}$$

5. The system is not stable, because if $x(t)$ is bounded, $y(t)$ still grows without bound because of the presence of the t^2 term

6. The system is memoryless. It is not time-invariant since

$$y(t - \tau) = (t - \tau)^2 \frac{dx(t - \tau)}{dt} \neq \hat{T}\{x(t - \tau)\}$$

7. The system is memoryless, causal, and stable. $\int_{-\infty}^{\infty} |h(\tau)| \, d\tau = 5$
8. e^{-2}
9. $y(t) = \frac{1}{5}(1 - e^{-5t})u(t)$
10. $\int_{-\infty}^{\infty} t^2 \delta(t + 5) \, dt = 25$

Chapter 3

1. $\pi/5$ rad/sample
2. $x_e[n] = (x[n] + x[-n])/2, \; x_o[n] = (x[n] - x[-n])/2$
3. $1/2$
4. $9/8$
5. The system is memoryless and causal
6. Yes, since $h[n] = 0$ for $n < 0$
7. No, since there are nonzero terms when $n \neq 0$
8. $y[n] = \{1, \; 3, \; 4, \; 3, \; 1\}$ where $0 \leq n \leq 4$
9. Yes
10. $u[n] - u[n - 3]$

Chapter 4

1. Use $b_n = (1/T_0) \int_{-T_0/2}^{T_0/2} x(t) \sin(2\pi nt / T_0)dt$
2. Use integration by parts, following the procedure in the example
3. $a_0 = a_1 = \cdots = a_n = 0, \; b_1 = 1/\pi, \; b_2 = -1/2\pi, \; b_3 = 1/3\pi$
4. $c_1 = 1/2 = c_{-1}, \; c_4 = 1/2j = c_4^*$, all other terms vanish
5. $j16\pi f/(-25 + 24\pi^2 f^2 - 16\pi^4 f^4)$
6. $2/(1 + 4\pi^2 f^2)$
7. $1/(1 + j2\pi f)$
8. $-j/\pi f$
9. $\frac{1}{2}e^{-j2\pi f}[-(j/\pi f) + \delta(f)]$
10. $e^{-j4\pi f}/(1 + j2\pi f)$

Chapter 5

1. $h(t) = 1/\pi t$
2. $y(t) = \sin(2\pi f_0 t)$
3. $Y(f) = 2e^{-j2\pi f}/j\pi f \Rightarrow y(t) = 2\,\text{sgn}\,(t-1)$
4. $u(t)(1 - e^{-t})$
5. The phase spectrum is shown below in Fig. C-1
6. $y(t) = (1/\pi)\sin \pi t - (1/2\pi)\sin 2\pi t$. See plot in Fig. C-2.
7. $X(\omega) = 1/(1 + j\omega)$, $E_X = 1/2$

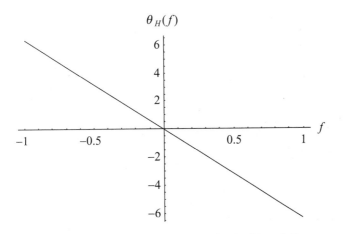

Fig. C-1. The phase spectrum for Problem 5-5.

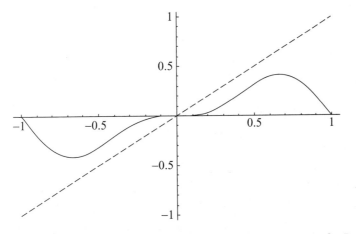

Fig. C-2. Dashed line shows original signal and solid line is filter output for Problem 6-5.

8. $Y(\omega) = \begin{cases} 1/(1+j\omega) & |\omega| < \omega_c \\ 0 & \text{otherwise} \end{cases}$

9. $E_Y = \arctan \omega_c/\pi$

10. $\omega_c = 1$ rad/s

Chapter 6

1. a
2. c
3. c
4. d
5. a

Chapter 7

1. $x[n] = \{5, 3, 4\}$, $N_0 = 3$, $\Omega_0 = 2\pi/3$
2. $c_0 = 4$, $c_1 = (1/2) + j(\sqrt{3}/6)$, $c_2 = (1/2) - j(\sqrt{3}/6)$
3. $X[k] = \{8, \ -3+j, \ -10, \ -3-j\}$
4. $x[n] = \{3, \ 0, \ 0, \ 1, \ -1, \ -1\}$.
5. The Fourier transform is

$$X(\Omega) = \frac{\sin[(7/2)\Omega]}{\sin(\Omega/2)}$$

The plot is shown in Fig. C-3.

6. $y[n] = \{1, \ 1, \ -1, \ -1\}$

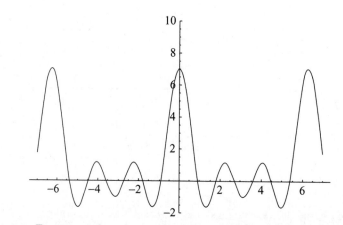

Fig. C-3. Frequency-domain representation of discrete-time square pulse.

Chapter 8

1. 9/2
2. No, at the minimum of the message signal $A_c + m(t) = 2 - 4 = -2 < 0$
3. $\eta = 33.3\%$
4. 0.31%
5. $m(t)\cos(\omega_c t)\sin(\omega_c t)\sin(\varphi t) = \dfrac{1}{2}m(t)\sin 2\omega_c t \sin(\varphi t)$. We can filter this term by rejecting ω_c
6. $y(t) = 0$

Chapter 9

1. The FM signal has a phase reversal of the lower sideband component
2. 3 MHz
3. $m(t) = (1/20)\sin(1000\pi t)$
4. $m(t) = (2t/\pi)\sin(1000\pi t) + (1000)t^2\cos(1000\pi t)$
5. $25.3(10^5)$ rad/s
6. 1.1 MHz

Chapter 10

1. $X(s) = 1$
2. $X(s) = s/(s^2 + \beta^2)$
3. $X(s) = \beta/(s^2 + \beta^2)$
4. $X(s) = 2\beta s/(s^2 + \beta^2)^2$
5. $y(t) = e^{-2t}(e^t - 1)$
6. $x(t) = [-(2/3)e^{-t} + (23/3)je^{2t}]ju(t)$
7. $x(t) = \delta(t - a)$
8. $h(t) = 3\delta(t) - 4je^{-2t}$
9. $h(t) = 4je^{-4t}u(t)$
10. $x(t)$

Chapter 11

1. $|z| > 1$
2. $X(z) = z\sin\omega/(z^2 - 2z\cos\omega + 1)$

3. $X(z) = z^{-m}$
4. $X(z) = z(z^2 + 4z + 1)/(z - 1)^4$
5. $x[n] = (u[n]/2)[1 - (-1)^n]$
6. $x[n] = n^2 u[n]$
7. $X(z) = 2 + 2z^{-1} + 2z^{-2}$
8. $\{1, -2, 6\}$
9. $\{1, 2, 3, 2, 1\}$
10. $y[n] = [(8/3) - 2(1/2)^n + (1/3)(1/4)^n]u[n]$

Chapter 12

1. The plot is given by

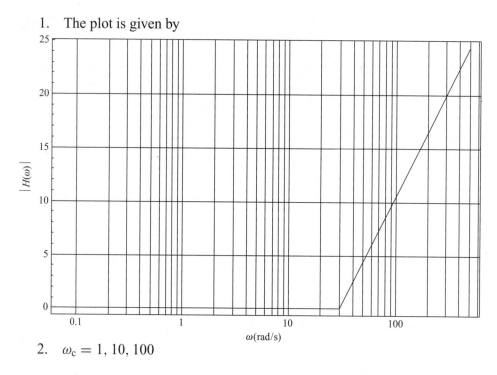

2. $\omega_c = 1, 10, 100$

3.

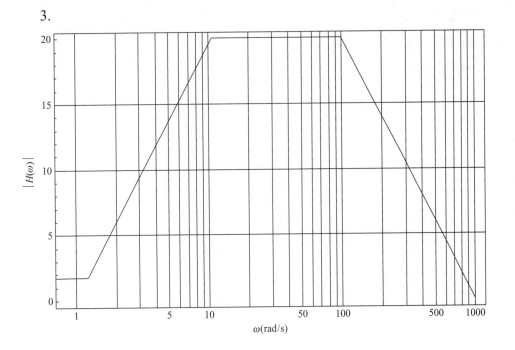

Final Exam
Answer Key

1. c	2. b	3. a	4. d	5. b
6. a	7. b	8. c	9. c	10. d
11. b	12. a	13. b	14. c	15. a
16. b	17. c	18. c	19. a	20. b
21. a	22. c	23. a	24. b	25. d
26. a	27. c	28. b	29. a	30. d
31. a	32. d	33. a	34. b	35. d
36. a	37. b	38. c	39. a	40. a
41. c	42. b	43. a	44. c	45. b
46. a	47. c	48. a	49. b	50. b
51. a	52. c	53. d	54. a	55. b
56. a	57. c	58. a	59. d	60. d
61. a	62. b	63. a	64. c	65. d
66. a	67. b	68. a	69. c	70. b
71. a	72. a	73. c	74. b	75. a

76. a	77. c	78. a	79. b	80. d
81. a	82. d	83. b	84. a	85. d
86. c	87. a	88. c	89. a	90. b
91. b	92. d	93. a	94. b	95. a
96. d	97. a	98. b	99. a	100. d

BIBLIOGRAPHY

This manuscript relied heavily on the following references, which the student will find useful after mastering the content of this book.

Briggs, W. and Henson, V.E., *The DFT: An Owners Manual for the Discrete Fourier Transform*, SIAM, Philidelphia, 1995.

Hsu, H.P., *Schaum's Outline of Signals and Systems*, McGraw-Hill, New York, 1995a.

Hsu, H.P., *Schaum's Outline of Analog and Digital Communications*, McGraw-Hill, New York, 1995b.

Haykin, S., *An Introduction to Analog and Digital Communications*, John Wiley & Sons, New York, 1989.

Roberts, M.J., *Signals and Systems: Analysis of Signals Through Linear Systems*, McGraw Hill, New York, 2003.

INDEX